MW00674695

THE LANGUAGE OF WORK

Technical Communication at Lukens Steel, 1810 to 1925

Carol Siri Johnson
New Jersey Institute of Technology

Baywood's Technical Communications Series
Series Editor: CHARLES H. SIDES

Baywood Publishing Company, Inc.
AMITYVILLE, NEW YORK

Printed in the United States of America on acid-free recycled paper.

Baywood Publishing Company, Inc.
26 Austin Avenue
P.O. Box 337
Amityville, NY 11701
(800) 638-7819
E-mail: baywood@baywood.com
Web site: baywood.com

Library of Congress Catalog Number: 2008030884
ISBN 978-0-89503-384-0

Library of Congress Cataloging-in-Publication Data

Johnson, Carol Siri.
The language of work : technical communication at Lukens Steel, 1810 to 1925 / by Carol Siri Johnson.
p. cm. -- (Baywood's technical communications series)
Includes bibliographical references and index.
ISBN 978-0-89503-384-0 (cloth : alk. paper) 1. Communication of technical information--United States. 2. Lukens Steel Company--History. I. Title.

T10.63.A1J64 2008
601.4--dc22

2008030884

To my father,
who would have loved this book.

Table of Contents

List of Figures and Sources vi

Introduction 1

PART ONE: BACKGROUND

CHAPTER 1
The Evolution of Technical Communication in the American
Iron and Steel Industry 15

CHAPTER 2
The History of Lukens Steel (1810-1925) 37

PART TWO: ANALYSIS

CHAPTER 3
1810-1870: Prediscursive Technical Communication 59

CHAPTER 4
1870-1900: Record Keeping Paves the Way 75

CHAPTER 5
Lukens 1900-1915: An Explosion of Technical Communication. 107

CHAPTER 6
Lukens 1915-1925: The Union of Words and Work 153

Conclusion 189
Glossary 191
Index 195

List of Figures and Sources

COVER ART WORK

INTRODUCTION

Figure 1 – Courtesy Hagley Museum and Library, Acc. 50, B-311.
Figure 2 – Courtesy Hagley Museum and Library, Acc. 50, B-2023.

Chapter 1

Figure 1 – Courtesy Hagley Museum and Library, Acc. 0339.
Figure 2 – Courtesy Hagley Museum and Library, Acc. 0339.
Figure 3 – Courtesy Hagley Museum and Library, Acc. 1631.
Figure 4 – Courtesy North Jersey Highlands Historical Society.
Figure 5 – Vannoccio Biringuccio, *The Pirotechnia*, courtesy Dover Publications, 1990.
Figure 6 – Courtesy North Jersey Highlands Historical Society.
Figure 7 – Courtesy North Jersey Highlands Historical Society.
Figure 9 – Courtesy Rutgers Library Annex.
Figure 10 – Courtesy of Simmons-Boardman Publishing Corporation, 1944.

Chapter 2

Figure 1 – Courtesy Hagley Museum and Library, Acc. 50, Pictorial Collections.
Figure 2 – Courtesy Hagley Museum and Library, Acc. 50, Pictorial Collections.
Figure 3 – Courtesy Hagley Museum and Library, Acc. 50, Pictorial Collections.
Figure 4 – Adapted from Thomas Bradford's *A Comprehensive Atlas Geographical, Historical & Commercial,* courtesy Harold Cramer, http://www.mapsofpa.com/.
Figure 5 – Adapted from *Gray's Railroad and County Map of Pennsylvania*, courtesy Harold Cramer, http://www.mapsofpa.com/.
Figure 6 – Courtesy Hagley Museum and Library, Acc. 50, Pictorial Collections.

Figure 7 – Courtesy Hagley Museum and Library, Acc. 50, Pictorial Collections.
Figure 8 – Courtesy Hagley Museum and Library, Acc. 50, Pictorial Collections.
Figure 9 – Courtesy Hagley Museum and Library, Acc. 50, Pictorial Collections.

Chapter 3

Figure 1 – Courtesy Hagley Museum and Library, Acc. 50, B-2216.
Figure 2 – Courtesy Hagley Museum and Library, Acc. 50, Pictorial Collections.
Figure 3 – Courtesy Hagley Museum and Library, Acc. 50, Pictorial Collections.
Figure 4 – Courtesy Hagley Museum and Library, Acc. 50, V-26, V-53, V-67.
Figure 5 – Courtesy Hagley Museum and Library, Acc. 50, B-2217.
Figure 6 – Courtesy Hagley Museum and Library, Acc. 50, B-1483.
Figure 7 – Courtesy Hagley Museum and Library, Acc. 50, B-1483.
Figure 8 – Courtesy Hagley Museum and Library, Acc. 50, B-1483.
Figure 9 – Courtesy Hagley Museum and Library, Catalogue for Yawman and Erbe, 1905.
Figure 10 – Courtesy Hagley Museum and Library, Acc. 50, B-2216.

Chapter 4

Figure 1 – Courtesy Hagley Museum and Library, Acc. 50, Pictorial Collections.
Figure 2 – Courtesy Hagley Museum and Library, Acc. 50, Pictorial Collections.
Figure 3 – Courtesy Hagley Museum and Library, Acc. 50, Pictorial Collections.
Figure 4 – Courtesy Hagley Museum and Library, Acc. 50, Pictorial Collections.
Figure 5 – Courtesy Hagley Museum and Library, Acc. 50, V-257.
Figure 6 – Courtesy Hagley Museum and Library, Acc. 50, V-252.
Figure 7 – Courtesy Hagley Museum and Library, Acc. 50, B-270.
Figure 8 – Courtesy Hagley Museum and Library, Acc. 50, B-310.
Figure 9 – Courtesy Hagley Museum and Library, Acc. 50, V-274.
Figure 10 – Courtesy Hagley Museum and Library, Acc. 50, V-298.
Figure 11 – Courtesy Hagley Museum and Library, Acc. 50, V-299.
Figure 12 – Courtesy Hagley Museum and Library, Acc. 50, B-2052.
Figure 14 – Courtesy Hagley Museum and Library, Acc. 50, B-17.
Figure 15 – Courtesy Hagley Museum and Library, Acc. 50, B-1214.
Figure 16 – Courtesy Hagley Museum and Library, Acc. 50, Pictorial Collections.
Figure 17 – Courtesy Hagley Museum and Library, Acc. 50, B-1227.
Figure 18 – Courtesy Hagley Museum and Library, Acc. 50, B-1245.
Figure 19 – Courtesy Hagley Museum and Library, Acc. 50, B-1245.

Chapter 5

Figure 1 – Courtesy Hagley Museum and Library, Acc. 50, Pictorial Collections.
Figure 2 – Courtesy Hagley Museum and Library, Acc. 50, Pictorial Collections.

Figure 3 – Courtesy Hagley Museum and Library, Acc. 50, Pictorial Collections.
Figure 4 – Courtesy Hagley Museum and Library, Acc. 50, B-4.
Figure 5 – Courtesy Hagley Museum and Library, Acc. 50, B-4.
Figure 6 – Courtesy Hagley Museum and Library, Acc. 50, B-2215.
Figure 7 – Courtesy Hagley Museum and Library, Acc. 50, Pictorial Collections.
Figure 8 – Courtesy Hagley Museum and Library, Acc. 50, B-1970.
Figure 9 – Courtesy Hagley Museum and Library, Acc. 50, B-1970.
Figure 10 – Courtesy Hagley Museum and Library, Acc. 50, B-4.
Figure 11 – Courtesy Hagley Museum and Library, Acc. 50, B-4.
Figure 12 – Courtesy Hagley Museum and Library, Acc. 50, B-4.
Figure 13 – Courtesy Hagley Museum and Library, Acc. 50, B-4.
Figure 14 – Courtesy Hagley Museum and Library, Acc. 50, Pictorial Collections.
Figure 15 – Courtesy Hagley Museum and Library, Acc. 50, B-1965.
Figure 17 – Courtesy Hagley Museum and Library, Acc. 50, B-310.
Figure 18 – Courtesy Hagley Museum and Library, Acc. 50, B-310.
Figure 19 – Courtesy Hagley Museum and Library, Acc. 50, B-4.
Figure 20 – Courtesy Hagley Museum and Library, Acc. 50, B-4.
Figure 23 – Courtesy Hagley Museum and Library, Acc. 50, B-8.
Figure 24 – Courtesy Hagley Museum and Library, Acc. 50, B-8.
Figure 25 – Courtesy Hagley Museum and Library, Acc. 50, B-2002.
Figure 26 – Courtesy Hagley Museum and Library, Acc. 50, B-2002.
Figure 27 – Courtesy Hagley Museum and Library, Acc. 50, B-2217.

Chapter 6

Figure 1 – Courtesy Hagley Museum and Library, Acc. 50, Pictorial Collections.
Figure 2 – Courtesy Hagley Museum and Library, Acc. 50, Pictorial Collections.
Figure 3 – Courtesy Hagley Museum and Library, Acc. 50, Pictorial Collections.
Figure 4 – Courtesy Hagley Museum and Library, Acc. 50, Pictorial Collections.
Figure 5 – Courtesy Hagley Museum and Library, Acc. 50, Pictorial Collections.
Figure 6 – Courtesy Hagley Museum and Library, Acc. 50, Pictorial Collections.
Figure 7 – Courtesy Hagley Museum and Library, Acc. 50, Pictorial Collections.
Figure 8 – Courtesy Hagley Museum and Library, Acc. 50, Pictorial Collections.
Figure 9 – Courtesy Hagley Museum and Library, Acc. 50, B-2001.
Figure 10 – Courtesy Hagley Museum and Library, Acc. 50, B-2001.
Figure 11 – Courtesy Hagley Museum and Library, Acc. 50, B-1991.
Figure 12 – Courtesy Hagley Museum and Library, Acc. 50, B-2001.
Figure 13 – Courtesy Hagley Museum and Library, Acc. 50, B-2001.
Figure 14 – Courtesy Hagley Museum and Library, Acc. 50, B-2001.
Figure 15 – Courtesy Hagley Museum and Library, Acc. 50, B-2002.
Figure 16 – Courtesy Hagley Museum and Library, Acc. 50, Pictorial Collections.

Introduction

THEORY AND HISTORY OF TECHNICAL COMMUNICATION

Technical communication, a form of writing often overlooked in literary scholarship, affords us a unique view of the discourse environments that make up our world. It provides examples of complex verbal and visual interactions of real people in the context of daily working life, not divorced from the communities in which they reside. This book is an analysis of layers of communication within a single industry between the years 1810 and 1925, the time during which the amount of technical communication began to increase exponentially. Most of these documents do not have a beginning, a middle, or an end; they are instances, fragments, or parts of a larger whole. Some are fragments of ongoing conversations, some are attempts to record present physical realities, some are self-promotion, but most are the visual and verbal remains of complex problem solving. From an analysis of technical communication at Lukens Steel, we can see that the industrial revolution would not have been possible without the attendant—and intrinsic—evolution of complex technical communication. Technical communication is a language essential to work in the modern world.

TOWARD A STRATEGY OF READING DISCOURSE COMMUNITIES

Behind this study is the notion that great stories are to be found not only in fiction, but in ordinary, everyday writing. Technical communication is seldom accomplished within a vacuum; it is usually a part of a larger ongoing community of speakers and listeners, writers and readers. Discourse communities are multiple, overlapping, and interactive; they have many authors, both named and unnamed, and many readers, both assumed and accidental. Just as we read fiction for stories with human interest, we can read technical communication for multiple complex stories with densely layered meaning. This strategy of reading can be applied to a single industry (as it is in these pages), to medical literature, to pamphlets, sewing machine directions, corporate reports, train schedules, game instructions, online discourse—to just about every situation in which people communicate with text. The possibilities for analyzing social discourse communities within their context are limitless.

When technical communication is analyzed as part of a larger discourse environment, the reader needs to ask: Why was this document made? Who made it? Who used it? Where was it kept and how was it treated? What value did it hold in its own time and what value does it hold now? Such questions open the door into a rich textual world that has barely been noticed by traditional interpretative literary analyses. Reading in this fashion requires looking at the entire document—the physical properties, as well as words—the binding, the source, the illustrations, the media, and the condition. In short, it requires a multidisciplinary contextual analysis. This approach to interpreting technical writing can be called archaeological.

It is possible to analyze discourse environments in a variety of ways. For instance, an analysis could focus on a single subject across time or compare different subjects that exist at a single time. It is possible to analyze technical communication in an entire industry, a portion of an industry, or a specific instance of the industry. This book is an analysis of one instance (company) within the American iron and steel industry, Lukens Steel, that began in 1810 and is still operated by a global steel corporation today. Lukens Steel provides a good sample for analysis because it was managed by members of one family for over 150 years and they saved the majority of their papers.

The changes in American manufacturing can be seen in the changing discourse environment at Lukens Steel. The documents in the early years were mainly deeds, handwritten letters, and complex accounting books, which included daybooks, journals, and ledgers. After 1849 Lukens used letterbooks to record outgoing correspondence: freshly written letters were pressed onto damp tissue paper as they were written. Technical communication was sometimes included in the letters, such as product specifications, drawings, and discussion of faulty materials, but it was rare. It was an important step when the first record-keeping documents appeared on the factory floor in the 1890s. From that point on, in addition to the letterbooks used by management, there were records of steel output from the open-hearth and plate mills, records of incoming and outgoing material in railroad cars, records of defects, and other records, such as notebooks carried by the foremen as in the example seen below (Figure 1). After 1900 general written and visual literacy can be seen in a wider group of authors from the factory floor. Managers, workers, foremen, and business owners communicated by a series of notes on small paper sent in an intrafactory mail system. After the introduction of the typewriter and carbon paper, managers, foremen, and the owners communicated via a third party, the stenographer typist. From that point on the amount of technical communication grew exponentially.

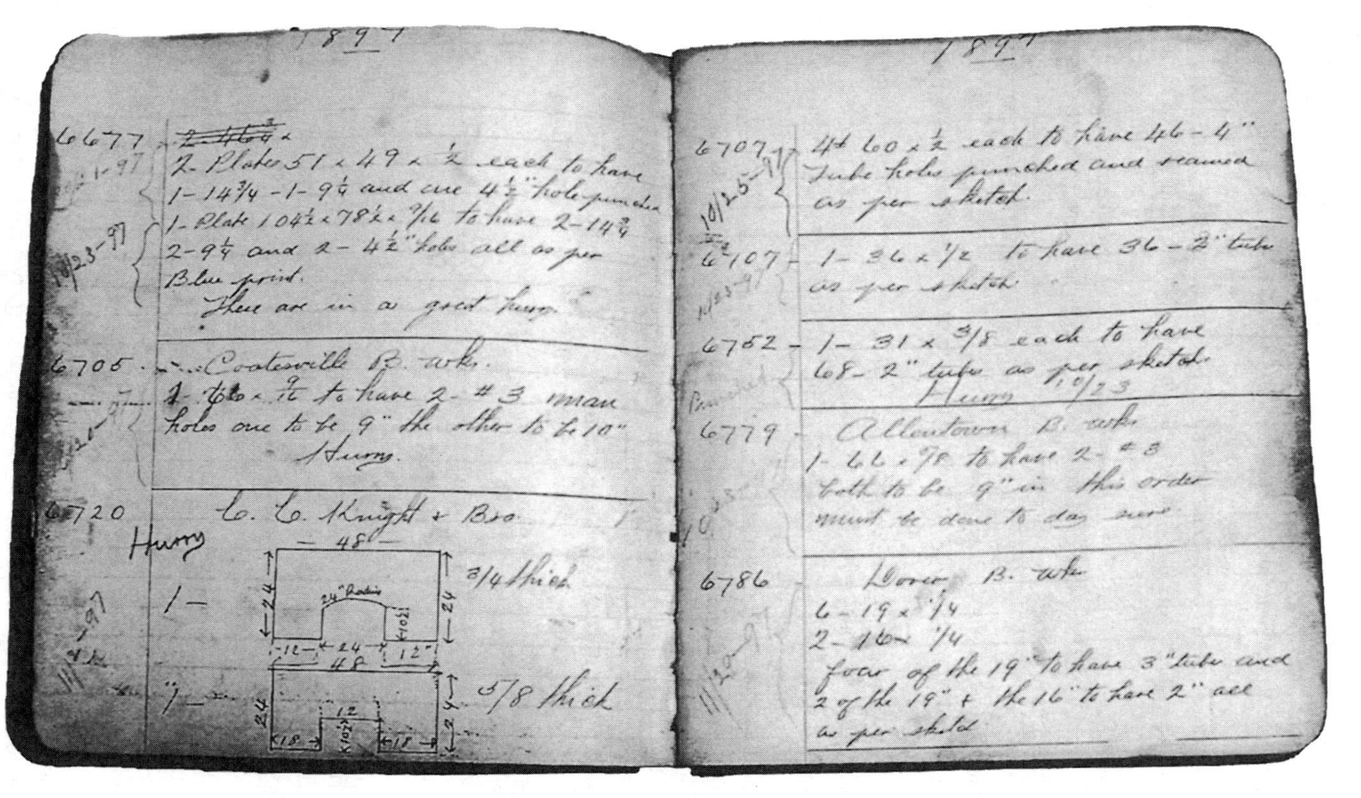

1897

6677 — 2-Plates 51 x 49 x 1/2 each to have 1-14 3/4 - 1-9 1/4 and one 4 1/2" hole punched

1-Plate 104 1/2 x 78 1/2 x 9/16 to have 2-14 3/4 2-9 1/4 and 2-4 1/2" holes all as per Blue print.

They are in a great hurry

6705 — Coatesville B. wks.

1- ... to have 2-#3 man holes one to be 9" the other to be 10"

Hurry.

6720 — C. C. Knight + Bro

Hurry

1- 48 ... 3/4 thick

1- ... 5/8 thick

1897

6707 — 4 ... 60 x 1/2 each to have 46-4" Tube holes punched and reamed as per sketch.

6707 — 1-36 x 1/2 to have 36-2" tube as per sketch

6752 — 1-31 x 3/8 each to have 68-2" tube as per sketch

Hurry 10/23

6779 — Allentown B. wks

1-66 x 7/8 to have 2-#3 both to be 9" in this order must be done to day sure

6786 — Dover B. wks

6-19 x 1/4

2-16 x 1/4

four of the 19" to have 3" tube and 2 of the 19" + the 16" to have 2" all as per sketch

Figure 1. **Punch Drill Order Book by William Malallieu (1897).** Textual analysis can merge with archaeology to help raise fundamental questions—What is the physical presence of the document? Where was it found? What status did it hold? How was it used in the ongoing dialogues, or social discourse community?

The theory behind my contextual method of reading was shaped by several texts, including the work of Elizabeth Tebeaux, JoAnne Yates, Charles Bazerman, and Michel Foucault's *The Archaeology of Knowledge*. This chapter briefly discusses how their ideas preceded mine. Tebeaux was the first to look at all texts, including instructional manuals, as being of literary value worthy of further study. Yates was the first to analyze the changing methods of communication in the corporate world. Bazerman described the evolving form of the scientific report and how it met the needs of a discourse community. Technical communication is far more important than scholars, to date, have realized, and these authors have recognized that. Foucault was influential in that he advocated reading the entire discourse community, within its context, and understanding that separate utterances, statements or authored texts do not exist without that which has gone before and that which comes after—all are parts of an ongoing conversation.

Technical communication is a method of knowledge exchange that reaches across both time and space; it enables people to codify received knowledge and helps to generate new knowledge. In the early days of the American iron industry, technical communication happened prediscursively, between people in immediate proximity, and thus knowledge traveled slowly, often embodied within a human being. In the nineteenth century, however, there was an explosion of communication, and for the first time knowledge could travel as fast as the railroads, newspapers, and journals. Since innovators could compare ideas and results more rapidly, this hastened the rate of individual inventions and the combination of inventions, eventually creating the technological world we live in today. The importance of technical communication should not be underestimated—it is the text of knowledge.

FOUCAULT: THE ARCHAEOLOGY OF KNOWLEDGE

In Foucault's "Archaeological Method and the Discourse of Science," Cynthia Haller writes that Foucault's method enables us to see "discourse not as a product of authorial intention but as a matrix within which relations of knowledge and power in society are created, maintained, and transformed . . ." [1, p. 56]. In the matrix, the warp and woof are connected so that power can't move without moving knowledge, and vice versa. Social power relations will shape modalities of technical communication at the same time that the capacity for technical communication will shape and modify social relations. During the time span of this book, specialized technological knowledge was at first embodied in individual people, but gradually it emerged as notations on paper. As management moved further away from the daily work and the complexity of that work increased, technical writing became necessary in most parts of the plant, including activities on the factory floor.

In *The Archaeology of Knowledge*, Foucault wrote that "history transforms *documents* into *monuments*" [2, p. 7]. In our society we have chosen to make what we call literature the primary genre for monumentalization. By doing this we memorialize an individual consciousness in a state of retrospection. The rupture offered by Foucault invites us to choose something else. Technical communication is a discourse form that is an external dialogue rather than an internal monologue, an attempt to communicate with multiple people for multiple purposes. It has a subject and a goal—the trading and creation of knowledge—and it is the literature of a group rather than an individual.

Rather than seeing a book as a simple unity, with an author, a beginning, a middle, and an end, Foucault reminds us that "The frontiers of a book are never clear-cut: beyond the title, the first lines, and the last full stop, beyond its internal configuration and its autonomous form, it is caught up in a system of references to other books, other texts, other sentences: it is a node within a network" [2, p. 23]. All of the boundaries that we observe represent choices that we make. What is deemed important by us at any given point in time differs. Interestingly, Foucault also reminds us that, for every emergence of a text or document, there is much, much more that has gone on behind the scenes to make it happen. He writes, "Behind the visible façade of the system, one posits the rich uncertainty of disorder; and beneath the thin surface of discourse, the whole mass of a largely silent development . . ." [2, p. 76]. Technical communication is especially interesting, because in it we can see a web of knowledge exchange that results in actions in the world. Thus in technical communication we can hear multiple, interweaving voices in networks of dialogue.

The basic building block of Foucault's *Archaeology of Knowledge* is the statement, or utterance. He defines a statement as "A seed that appears on the surface of a tissue of which it is the constituent element," or in other words, as part of a network: "There is no statement that does not presuppose others; there is no statement that is not surrounded by a field of coexistences . . ." [2, p. 88]. In the case of the American iron and steel industry, the statement may take many forms: a scrawled note, a list of defects, a test report, a few words passed between workers, a letter with drawings, or a calculation. It is these elements that make up the larger discourse environment.

A discourse is a group of statements formed by specific discursive practices. Discursive practices consist of the assumptions and rules underlying a discourse (such as the methods for filling out a certain type of test report). When you approach language in this manner, new worlds and possibilities open up. Reading documents as statements from a discourse environment allows many levels of interpretation: such a strategy makes it possible "to snatch past discourse from its inertia and, for a moment, to rediscover something of its lost vitality" [2, p. 107]. Foucault also notes that, within discourse communities, some topics are prohibited. In a speech delivered to the Collège de France, Foucault said (as translated), "I am supposing that in every society the production of discourse is at once controlled, selected, organized and redistributed according to a certain number of procedures . . . . The most obvious and familiar of these concerns what is prohibited. We know perfectly well that we are not free to say just anything . . ." [2, p. 216]. One of the prohibitions in the academic world today concerns the definition of "literature"; presently, it is defined as fiction, religious writing, poetry, and essays. There is a vast wealth of writing from the working world that can be studied in context and yet, because it is relatively new (expanding in the late nineteenth century), is not perhaps considered literature.

TEBEAUX: READING TECHNICAL COMMUNICATION

Elizabeth Tebeaux wrote the first book-length analysis about the role of technical communication in history in 1997, *The Emergence of a Tradition: Technical Writing in the English Renaissance, 1475-1640*. Using Pollard and Redgrave's *A Short-title Catalogue of Books Printed in England, Scotland, & Ireland and of English Books Printed Abroad, 1475-1640* as a source, she discovered hundreds of texts that have not been read or analyzed as part of the modern canon. She argued that "technical writing, like literature, history, and philosophy, is worthy of study in its own right" [3, p. 3]. She notes that studies of the English Renaissance have long been focused on "courtiers, drama, political intrigue, political and theological polemic, military and geographical conquests, love poems, catechisms, sermons and worship aids" and she suggested that reading technical writing provides "scholars of language a broader understanding of the characteristics of a period than literary or historical studies alone will afford" [3, p. 2]. In addition, reading technical communication as a literature within an everyday context can provide an unfiltered, contemporaneous glimpse of people working together in ordinary life.

Very little attention has been paid to writing about working people or ordinary working activities, such as "farming, gardening, animal husbandry, surveying, navigation, military science, accounting, recreation, estate management, household management, cooking, medicine, beekeeping, and silkworm production, to name a few," even though many of these documents are widely available [3, p. 3]. Tebeaux theorizes that little attention is paid to technical communication as literature because it bears "the taint of the marketplace and the non-academic world" [3, p. 3]. This is an example of how, as Foucault describes, discourse communities create their own rules of formation and prohibit areas of inquiry. Perhaps it's time to expand the boundaries of what we read and what we teach to have a richer and more accurate vision of the world.

Technical communication takes many different forms, which scholars have hardly even begun to address. Each type and instance of technical communication is part of an ongoing dialogue, an interconnecting, overlapping, responsive network with knowledge moving from node to node. Moreover, since examples of technical communication frequently incorporate many visual elements, their analysis spills over into the field of graphic art. Layout is used to guide the eye, and drawings, woodcuts, photographs, tables, and graphs are used, in tandem with words, to attempt to communicate physical meaning. In 1983 John Brockmann noted that the history of technical communication had, until that point, focused mainly on great authors or scientists. This myth of the "genius in the attic" has long been a part of our culture. In reality, inventions emerge from multiple minds, and they emerge more rapidly when the minds are in communication: new technology develops cumulatively by a series of inventions, modifications and communications, rather than by a single person at a single time [4, p. 19]. We are gradually moving away from the myth of the heroic genius and toward a complex and nuanced view of the world and its functioning, as being composed of many people with many voices and visions. Brockmann suggested examining a "broad spectrum of writers," including those who are "uncelebrated," since it would be "immensely more accurate and meaningful" [5, p. 156]. It is time to look at the whole, rather than the part.

YATES: BEYOND INDIVIDUAL MEMORY

JoAnne Yates's *Control Through Communication: The Rise of System in American Management* brought business communication, another subset of transactional rhetoric, to the fore as a discourse worthy of analysis. Like Tebeaux, Yates is a pioneer who saw the possibilities in studying a form of communication that has, thus far, been overlooked by the majority of scholars: corporate communication. Yates took as her topic the evolution of corporate communication in the railroad and manufacturing industries during the nineteenth and early twentieth centuries. Just as Tebeaux outlines the history of how printing contributed to the explosion of knowledge in the fifteenth century, Yates examines how the railroad and telegraph changed the face of both business and communication. For instance, railroads were central to the evolution of business communication: they first required and then defined some of the fundamental underlying methods of modern corporate communication. Yates argues that the foundation of the modern corporation was created by the bureaucracy necessitated by railroads, which required accurate time tables and notification of accidents, as well as other exact data.

> Before 1850, the economy was dominated by small firms owned and managed by a single individual or a partnership and operating in a local or regional market. The spread of the telegraph and of railroads around the middle of the century encouraged firms to serve larger, regional and national markets, while improvements in manufacturing technology created potential economies of scale [6, p. 1].

Through the growth of rapid travel and communication, larger networks of business and industry were possible. These changes—the overcoming of geographical boundaries—were essential in making possible the technologically advanced society of today. The new distributed corporations were woven together with communication: for the first time, communication was necessary, on a daily basis, between all levels in the hierarchy of workers. "Only through such communication could managers have any hope of coordinating the many physically separated individuals and activities required to make the modern corporation work" [6, p. xi]. The structure that emerged in the railroads gradually spread to other industries. A large part of Carnegie's success in the steel industry was that he and his partners had originally worked in the railroad and telegraph industries, so they knew how to make use of this web of communications that could distribute texts and hold a large group together across time and space.

The increase in business communication is parallel to the increase in technical communication, and the genres overlap: at times technical material was included within business communication, and some business documents required technical information. The methods of communication in the two genres overlap as well: Yates described the telegraph, typewriter, duplicating methods, letterbooks, and filing systems, all of which were essential to the emergence of modern corporations. This communication was necessary to overcome the limitations of the individual memory: "The published rules, the journal of operations, and the monthly reports all reflected a desire to rise above the individual memory and to establish an organizational memory tied to job positions and functions, rather than to specific individuals" [6, p. 6]. In both cases the organizational and technical communication had to "rise above the individual memory." The creation and continuation of business and industry had become a group act rather than an individual one. Communication that had been used in small groups and family businesses became insufficient as the complexity of technology grew beyond the capacity of the individual mind. This is another reason for studying community, rather than individual, communications.

BAZERMAN: THE EVOLUTION OF SCIENTIFIC DISCOURSE

Scientific discourse is another form of communication, one which uses structure and precedent to arrive at discovery and consensus. Charles Bazerman approached scientific discourse from both archaeological and anthropological standpoints. He wrote about viewing "text as a historical event within the unfolding context" [7, p. 3]. Each member of a scientific discourse "writes as part of an evolving discussion, with its own goals, issues, terms, arguments, and dialect" [7, p. 5]. This viewpoint parallels Foucault's archaeological theory of the relativity of texts and the evolving strictures of social discourse communities.

Like Yates and Tebeaux, Bazerman wrote about the power inherent within text. Yates wrote about communication creating structures of social power, and Tebeaux wrote about the knowledge dissemination that gave ordinary people more power over their lives. Bazerman noted that scientific writing has given us "increasingly immense control of the material world in which we reside" [7, p. 13]. Technical communication is embedded within a structure and creates power for its users. Massive iron and steel production would never have been possible without increased communication, including the interaction with the emerging fields of chemistry, testing, and standards that gave us enough power to build the complex infrastructure we use today.

Like the business communication that Yates describes, technical communication evolved and shifted in genre in order to fill a variety of needs. Words for many of the elements in the iron industry, such as the word "steel," were continually being negotiated and redefined. None of this discourse sprang fully formed from the head of Zeus. The same was true of the development of scientific discourse. Bazerman argues that "Symbolic systems react to experiences and situations, to contact with different communities and the formation of new communities, to struggles with old meanings deemed inadequate to account for emerging ideas and experiences, to the need to create shared understanding and agreement where none existed previously" [7, p. 21]. In his analysis of scientific articles, Bazerman found that the gradual emergence of its form was a product of consensus and that each article was the product of consensus as well [7, pp. 22-23].

The scientific article came to represent a distinct genre of writing and communication. In business and technical communication, forms shift continually according to need. Carolyn Miller defined genres as "typified rhetorical actions based in recurrent situations" that can lead to successful group action [8, p. 159]. Different forms of writing emerge and evolve to fill different needs. In the case of Lukens Steel, several new forms emerged at the turn of the nineteenth century, including intensive reporting of data, standardized forms, multiple copies of drawings, testing reports, and written technical communication between the owners, managers, foremen, and workers. Later, complex typewritten letters and reports created a new class of worker, the stenographer typist, who became essential to industry as well.

OTHER STUDIES IN THE HISTORY OF TECHNICAL COMMUNICATION

Brockmann wrote that studies of the history of technical communication until 1983 had mainly reflected the "great author" method of discourse analysis [5, p. 155]. The first articles about the history of technical communication often focused on science as well. For instance, in 1960 Joel Shulman wrote "Technical Writers Who Became Famous as Scientists" and in 1961 Charles Hargis wrote "America's First Great Technical Writer" naming Benjamin Franklin [9, p. 10]. In the 36 articles in the 1983 Brockmann bibliography, about half refer to generic, institutional, or group matters and the other half are devoted to individual authors, especially Chaucer and Franklin. Examples of individual technical writers writing in isolation, such as Leonardo Da Vinci, can be studied as technical communication, but individual geniuses are the exception rather than the rule: they exist outside the context of working people in the working world. The reward for looking at technical communication within its context is seeing what happens in the a human network rather than an individual mind.

* * * * *

Louise Rosenblatt was the first to attempt to "broaden the framework" of literature in *Literature as Exploration* in 1938. Rosenblatt distinguished between *aesthetic* reading (predominantly literary) and *efferent* reading (predominantly nonliterary). Aesthetic reading is focused on the personal whereas efferent reading is part of a group dialogue that seeks to be read for a specific purpose [11, p. xvii]. She pointed out that a literary work contains "a special kind of intense and ordered experience—sensuous, intellectual, emotional—out of which social insights may arise" [11, p. 31]. She broadened our reading of literary work to include the reader's response, and perhaps it is time to broaden it further to include that which has been historically called nonliterary. Technical communication is also about the senses and the intellect. What it lacks in emotion, it gains from its rich context, since it is almost always part of an ongoing group discourse.

Can technical communication be read as literature? According to Terry Eagleton, "Some texts are born literary, some achieve literariness, and some have literariness thrust upon them" [12, p. 7]. What we consider to be literature is our choice. The technical communication in the American iron and steel industry is multilayered, dialogic, and full of examples of writing, drawing, research, changing social relations, developing genres, and the growth of an industry. In this analysis of Lukens Steel, we can see an extraordinary company continually repositioning itself to survive. In fact, they stayed in business until 1998, and part of the plant is still being operated by a global steel corporation today. Lukens Steel was a fusion of family and worker, of intellectual inquiry and chaotic structure, of opposites and ironies, as can be seen in a painting of Rebecca Lukens on a military repair ship in 1947 (Figure 2). Other companies have their own characteristics as well and can tell different tales. The documentation that they produced, if they saved it, is a fluid conversation caught in time. It is a different sort of literature than a novel, but it still tells the story of human lives.

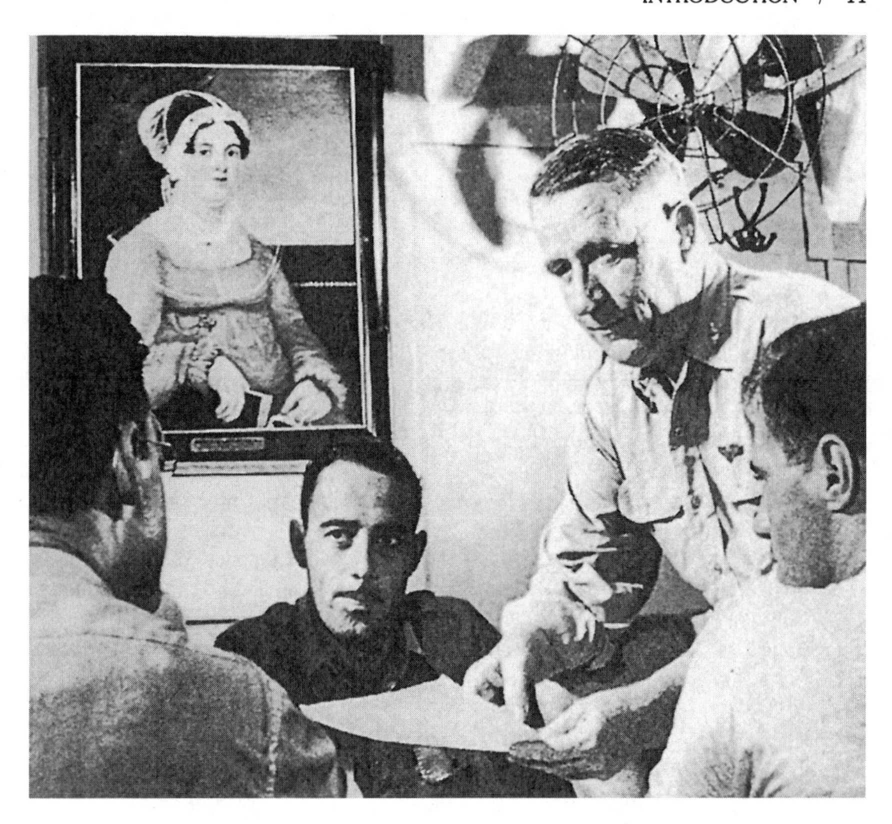

Figure 2. **Photo from Floating Aircraft Repair Unit, United States Army Air Force Special Services "Rebecca Lukens" (1947).** A nineteenth century painting of Rebecca Lukens as a backdrop for a repair ship that carried machine shops for precision work during World War II [13].

REFERENCES

1. C. R. Haller, Foucault's Archaeological Method and the Discourse of Science: Plotting Enunciative Fields, in *Essays in the Study of Scientific Discourse: Methods, Practice, and Pedagogy*, J. T. Battalio (ed.), Ablex Publishing, Stamford, Connecticut, pp. 53-69, 1998.
2. M. Foucault, *The Archaeology of Knowledge and the Discourse on Language*, Pantheon Books, New York, 1972.
3. E. Tebeaux, *The Emergence of a Tradition: Technical Writing in the English Renaissance, 1475-1640*, Baywood, Amityville, New York, 1997.
4. P. B. Meyer, "Episodes of Collective Invention," Working Paper 368, *United States Department of Labor*, Washington, D.C., August 2003, accessed Nov. 11, 2007 from http://opensource.mit.edu/papers/meyer.pdf

5. R. J. Brockmann, Bibliography of Articles on the History of Technical Writing, *Journal of Technical Writing and Communication, 13*:2, pp. 155-165, 1983.
6. J. Yates, *Control through Communication: The Rise of System in American Management*, Johns Hopkins University Press, Baltimore, Maryland, 1989.
7. C. Bazerman, *Shaping Written Knowledge: The Genre and Activity of the Experimental Article in Science*, University of Wisconsin Press, Madison, Wisconsin, 1988.
8. C. R. Miller, Genre as Social Action, *Quarterly Journal of Speech, 70,* pp. 151-167, May 1984.
9. J. J. Shulman, Technical Writers Who Become Famous as Scientists, *Society of Technical Writers and Publishers, 7*:2, pp. 17-21, 1960.
10. C. C. J. Hargis, America's First Great Technical Writer, *Society of Technical Writers and Publishers, 8*:4, pp. 14-15, 1961.
11. L. M. Rosenblatt, *Literature as Exploration* (5th Edition), Modern Language Association, New York, 1995.
12. T. Eagleton, *Literary Theory: An Introduction* (2nd Edition), University of Minnesota Press, Minneapolis, Minnesota, 1996.
13. J. S. Huston, "General Correspondence, Lukens Steel Company History, Rebecca Lukens," B-2023, *Lukens Steel Archives*, Hagley Museum and Library, Wilmington, Delaware, 1925-1948.

PART ONE

Background

CHAPTER 1

The Evolution of Technical Communication in the American Iron and Steel Industry

Technical communication has always been with us. Although its widespread use is a recent phenomena, technical texts exist wherever writing exists. From the documents left by Philo of Byzantium (c. 280-220 BCE) through Arabic works in the Persian Empire to medieval and Renaissance texts, authors have been writing about technology for thousands of years. In the iron industry, technical communication can be seen in fragments from the Greek and Roman Empires [1]. Al-Kindi, Abu Yusuf Ya'qub wrote treatises on mineralogy and metallurgy in the ninth century CE [2, p. 22]. Renaissance texts were the first blossoming of technical communication about mining and metallurgy, but it wasn't until approximately 1850 that the texts moved out of the sphere of intellectual scribes and into the working world. The use of technical communication, both writing and drawing, to exchange information grew with the industrial revolution: technology became so complex that we had to "rise above the individual memory and to establish an organizational memory" [3, p. 6].

Prior to 1850 most technical information was transmitted orally and by immediate physical proximity (watching and imitating). The Renaissance texts were available to only a few, and even if they had been available, "Books published before 1800 offered little practical advice; they were unlikely to have been of much use to working ironmakers" [4, p. 313]. The fragments, or incidental technical communication discussed later, are exceptions rather than the rule, and would have reached a small audience, if any. The changes that are documented in this book occurred when large numbers of people from different groups began using writing as a method of problem solving and creating new technology. At first these were working ironmasters, but as time went on the Venn Diagram (overlapping spheres) of foremen, workers, managers, engineers, and outside experts who communicated via writing and drawing grew until it could not be named or contained. By 1925 technical writing had become so fully integrated within our culture that the volume of our written interaction is nearly infinite. We have moved from an orally-based culture to one predicated on chirographic methods of knowledge creation and transfer. We use writing and graphics as an aid to thinking, and written documents are statements in conversation with each other.

Still, the earliest works on metallurgy in America were incidental or accidental texts. The first official and consistent documents were furnace journals which, although they mainly contained accounting records, were fluid in form and sometimes had technical information as well. In the late nineteenth century, however, publications surrounding the manufacture of iron and steel exploded with writing as experts exchanged knowledge at a rate hitherto unknown: state reports, commercial books, transactions of professional associations, and illustrated trade newspapers flooded the market. These publications were an outgrowth of the increase in technical communication that both accompanied and enabled rapid invention, modification of invention, and its wide and varied implementation. This chapter will describe the background of technical communication in America against which we can see the specific case of Lukens Steel, wherein technical communication moved from the purview of the owner-operators to the workers, foremen, and managers at roughly the same time.

EARLY MINING AND METALWORKING TEXTS

The earliest authors of technical documents were military engineers. Writers such as Philo recorded the earlier experiments of Ctesibius regarding catapults, pneumatics, and fortresses, for example, in the third century BC [5, p. 25]. Although these writings are not about mining and metalworking, they set a precedent. Sources from this time period are difficult to study because they are "usually poorly preserved and often frankly bewildering" [6]. Similarly, the technical texts from the Persian Empire are written in Arabic and often untranslated and uncatalogued. During the Middle Ages, secrecy and magic limited the open communication of knowledge, but with the Renaissance and the invention of the printing press, many books were published about technical communication. In the sixteenth century, two illustrated books were published about metalworking: Agricola's *De Re Metallica* (1556) and Vannoccio Biringuccio's *The Pirotechnia* (1540). Although the illustrations in these books were doubtlessly useful for readers seeking information, they were probably not used in the workplace. The former was published in Latin, which had no corresponding words for mining and metallurgical processes [7, p. iii]. The latter was written in Italian and translated into other languages, but few ironworkers would ever have seen or even have known about these volumes. The spheres of writing and work were still widely separated: these documents were collections of received knowledge rather than a medium of exchange.

INCIDENTAL TECHNICAL COMMUNICATION IN THE EARLY AMERICAN IRON INDUSTRY

William Byrd

Nearly as soon as America was settled, investors attempted to use its resources to produce iron. One of the earliest written records of how ironworks operated is the personal journal of William Byrd (1674-1744), in the section called "A Progress To The Mines" [8]. Byrd wrote this journal as notes to himself and his family and friends between the years 1728 and 1736. In 1841 it was published as a historical resource, and thus we have a window into early methods of knowledge exchange—at that point, the only way to find out how to do something was to find a person doing it and ask questions.

Byrd traveled through Virginia, near the Chesapeake Bay, and saw several ironworks, recounting his experiences in his journal. In the early colonial time period, taverns were few and far between, so travelers often stayed at local homes. Much of his journal consists of descriptions of his hosts (many of whom were women), the food, and weather. The following demonstrates his method of gathering information about the iron industry. Mr. Chiswell's ironworks was near the Pamunky River in Virginia. First, Byrd describes the reaction to his initial inquiry:

> I found Mr. Chiswell a sensible, well-bred man, and very frank in communicating his knowledge in the mystery of making iron, wherein he has had long experience. I told him I was come to spy the land, and inform myself of the expense of carrying on an iron work with effect. That I sought my instruction from him, who understood the whole mystery, having gained full experience in every part of it; only I was very sorry he had bought that experience so dear.

Chiswell was very willing to share his knowledge and took Bryd on a tour of his works. He gave a great deal of advice, which Byrd recorded. He described the necessity of assessing the ore, the importance of water power, and woodland for charcoal fuel. Also, in the American south the economy was based on slave labor, so there had to be land sufficient to grow corn. He summarized that "if all these circumstances should happily concur, and you could procure honest colliers and firemen, which will be difficult to do, you may easily run eight hundred tons of sow iron a year [8, pp. 127-128]. When Byrd wrote that Chiswell had "bought his experience so dear," it was because the ironworks were standing idle for lack of water. The variables that went into the process of making iron were so numerous that many of the ironworking operations in early America failed. The major reasons were poor waterpower, too great a distance to market, insufficient ore, or not enough workers.

Although Byrd's "A Progress To The Mines" contained technical information about the early iron industry, it was a private journal. He had no intention of publishing it, and it was not published during his lifetime. Thus, it was not written for the purposes of knowledge exchange, and it was not in dialogue with other documents. Many of the early ironworks in the Colonies failed since ironmaking knowledge was still embedded in the workers—as Chiswell said, "honest colliers and firemen, which will be difficult to [find]" and knowledge moved only when the workers moved.

Tuball Ironworks

Later, Byrd visited another troubled ironworks that belonged to the Governor of Virginia, Alexander Spotswood (1676-1740). By 1739 Spotswood was trying to rent his ironworks, leaving another incidental document about the American iron industry: he published *Iron Works At Tuball*, a formal description of the ironworks, terms, and conditions that were later made into a facsimile by the University of Virginia [9]. Spotswood gained ironworking knowledge by importing German workers who embodied it. However, many of the indentured servants left as soon as their time was up in order to seek out new opportunities. Thus, he came to rely on the use of slaves who had become highly skilled workers [9, p. 15].

During this time, the only consistently successful ironworks were the Principio and Baltimore Ironworks, both of which processed bog iron near major water routes. When Byrd visited Washington's ironworks on the Potomac, he described the manager, Mr. England, who "can neither write nor read; but without those helps, is so well skilled in iron works, that he does not only carry on his furnace, but has likewise the chief management of the works at Principia, at the head of the bay" [8, pp. 138-139]. Thus, technical knowledge about ironmaking in eighteenth century America was tacit knowledge embodied within an individual. The operations were small enough that there was no need for written communication to help them do their work.

Swedish Industrial Spying

Another type of incidental writing about the early American iron industry was *Report about the Mines in the United States of America, 1783* by Samuel Gustaf Hermelin (1744-1820), an industrial spy from Sweden. He was working within a tradition of industrial spying—Emanuel Swedenborg also wrote *De Ferro* [10, p. 9]. In the eighteenth century, England had run out of charcoal fuel and had begun importing iron from Sweden, thus making Sweden a crucial supplier. Hermelin was sent by the king to assess the status of their competitor. His account is mainly a narrative listing the existing ironworks in the United States, with a comparative analysis of the cost of labor and product. He describes some of the geological strata in Pennsylvania, the ores, the workers, and the industrial processes. His conclusion, at the beginning of the report, is that "In some of the United States of America, rich iron ores are obtained at cheap prices on account of the nature of the ore quarries, but, on the other hand, considerable expenses are incurred in manufacture through [the existing] high coal prices and wages" [10, p. 15]. Thus, American iron was not a significant threat to the Swedish iron industry because the power of the worker, sole repository of knowledge, allowed them to demand high wages.

Hasenclever's Apologia

Peter Hasenclever (1716-1793) provided another example of the power of the worker as the sole carrier of ironmaking knowledge. His apologia, *The Remarkable Case of Peter Hasenclever, Merchant*, describes the distributed ironworking empire that he built in 1764. To import knowledge, Hasenclever hired experienced laborers from Germany—forgemen, furnace men, charcoal burners, miners, masons, and carpenters—and transported them, along with their families, to New Jersey. These experienced workers built five ironmaking villages, each with a furnace, multiple forges with multiple fires, stamping mills, coal houses, blacksmith shops, houses, saw mills, reservoirs, ponds, bridges and roads, in an extremely short time, from 1764 to 1767. However, since wages were high and workers scarce, Hasenclever's workers quickly became contentious. Despite the language barriers, they learned from native workers that Hasenclever's wages were low, so they demanded more money. This, combined with Hasenclever's overspending (nearly £40,000 in three years) resulted in his being recalled to England and eventually declaring bankruptcy [11]. This was the source of the apologia, which, like Byrd's journal, Spotswood's prospectus, and Hermelin's report, was not written to communicate technical knowledge. These texts were written for other, incidental purposes; technical knowledge was still embodied within the worker.

Robert Erskine's Letters

In 1771 Robert Erskine (1735-1780) came to America to attempt to save Hasenclever's failing ironworks. However, since Erskine was trained as a hydraulics engineer, he had to quickly learn about the iron industry before he set out. Thus, like Byrd, he traveled: he went through England, Wales, and Scotland and took detailed notes of everything he saw, sending them as letters to his employer. Before he left for America, he retrieved the letters and used them as technical notes written to himself. These letters are some of the most clear and detailed descriptions of the existing ironworks and ironworking processes that exist from this period. The letters describe, as exactly as possible, blast furnaces, blowing engines, forges, foundries, steel works, processes, and experiments. As an example, he described an early steam engine, a technology that would not come to be used extensively in America for another half century:

> . . . the Bellows is an Iron Cylinder about 6 feet diam[r]. Worked by a fire Engine: The stroke is between 5 + 6 feet; the cylinder is fixed, and the piston enters at the bottom, and Works upwards, the piston rod from the beam of the fire Engine, branching into four rods, which descend on the outside of the Cylinder. When the piston rod ascends the air escapes through two tubes at the Top; but were such a body of air confined to the pipes which conveyed it to the furnace, some part of the machine must give way, and the blast would likewise intermit; the Tubes from the Cylinder therefore communicate with two Cast Iron Vessels about 8 feet Diam[r]. Inverted in Cisterns of Water, very much loaded to prevent their being buoyed up; in these, the Air, by forcing out the Water, expands itself, and the pressure of the water returning into them, continues the Blast, when the piston is going down; the Air returns through the same pipe it entered the Air vessels, and as Valves prevent its again entering the Cylinder, it passes through a square Tube made of Wood, to the furnace [12].

This is an example of fine technical writing; but, since the letters were written to gain personal knowledge and not part of a social dialogue, they were still incidental to the industry as a whole. Like Byrd, Erskine was using writing to create knowledge for an individual, not as part of a group dialogue. Erskine, too, found that the real ironmaking knowledge resided within the workers and it was to the workers, that he addressed the majority of his questions [13, p. 176].

There are myriad examples of such incidental technical communication surrounding not only the iron industry but every industry. The consistent and widespread use of written technical communication in the workplace did not begin until the nineteenth century, and then the forms that emerged were additive—the earlier forms were kept and more types joined them in the fluid discourse of ongoing information exchange. For instance, we still use prediscursive, tacit communication in which people learn directly from each other; there is no better method of knowledge exchange than sharing multiple sensory experiences in close physical proximity. Thus, even as written abstractions became a tool with which we modify the world, the oldest method of knowledge transfer—that of watching and imitating—is still fundamental in the working world. Later, the addition of writing and drawing as methods of knowledge transfer allowed more minds to interact, theorize, and create new ideas.

FURNACE JOURNALS

Furnace journals were the fundamental written records that served the ironmaking industry during the seventeenth, eighteenth, and nineteenth centuries. They are a rich source of knowledge for economists and historians because life can be reconstructed from them. For instance, the historian Charles Dew wrote *Ironmaker to the Confederacy* (1966, 1999) and *Bond of Iron: Master and Slave at Buffalo Forge* (1994) from a series of furnace journals and housekeeping books that he found scattered throughout the south and the Midwest; he was able to reconstruct daily life from these journals [14]. Most furnace journals were kept for accounting. Elizabeth Tebeaux wrote about how accounting methods evolved from a single book to "several books, each serving a different purpose [15, p. 319]. Principio had hundreds of books. Most ironworks had at least three—a day book, a journal, and a ledger. These documents are interesting because they gradually evolved to serve other needs, such as recording incoming supplies, outgoing products, shipping charges, and even the life of a town. At Lukens Steel they were the predecessor of the record keeping described in Chapter 4.

Day books, also called waste books, were usually written in every day, recording events as they happened; journals were intermediary books between day books and ledgers, which consolidated the chronological events under subject name headings; ledgers summarized the financial data for specific accounts and were usually updated once per year. The three books were often cross-indexed with numbers representing subject names, and the ledgers had an index as well. Since legal tender was scarce in early America, the accounts were seldom settled—just noted—and ran on from year to year. Many other books were used as well, such as collier's books for keeping information about charcoal, blast books about the operation and output of the furnaces, records about the company's store, and time books for workers. Many of the smaller ironworks kept all of the above information in one book, leaving space between sections.

Furnace journals were the first writing kept and used at the ironworks and, as such, they gradually developed into other genres. The Martha Furnace Diary and Journal (1808-1815) kept daily information, such as the weather, the operating of the furnace, fights, injuries, and drunkenness on the right page, and time information about the workers on the left (Figures 1 and 2). Some entries are mundane, such as "Owen Hedger quit. Jeremiah Fealdon took his team & boarding in the kitchen," and some have a wry humor, such as "The general topic of conversation here is respecting a dog that passed by today. It was said that he bit several dogs and was raving mad & some say that he yet retains his Mental Faculties" [16, p. 53].

George Nock His Book Ramapo Works (1837) evolved from a typical accounting furnace journal into incidental technical communication. This small journal began with standard record keeping data, but soon the author abandoned this material and filled it with recipes for making different types of iron and steel as well as descriptions of other technological processes (Figure 3) [17]. Throughout the world, accounting practices predated other forms of writing, and it is visible in the iron industry as well: the Nock journal shows a shift from financial record keeping to early technical writing. Prior to the nineteenth century, thinking about and creating new ironmaking technology happened prediscursively, with workers in close proximity sharing knowledge. As time went on, the more abstract act of writing enabled people, such as George Nock, to take the tacit knowledge and write it down on a page.

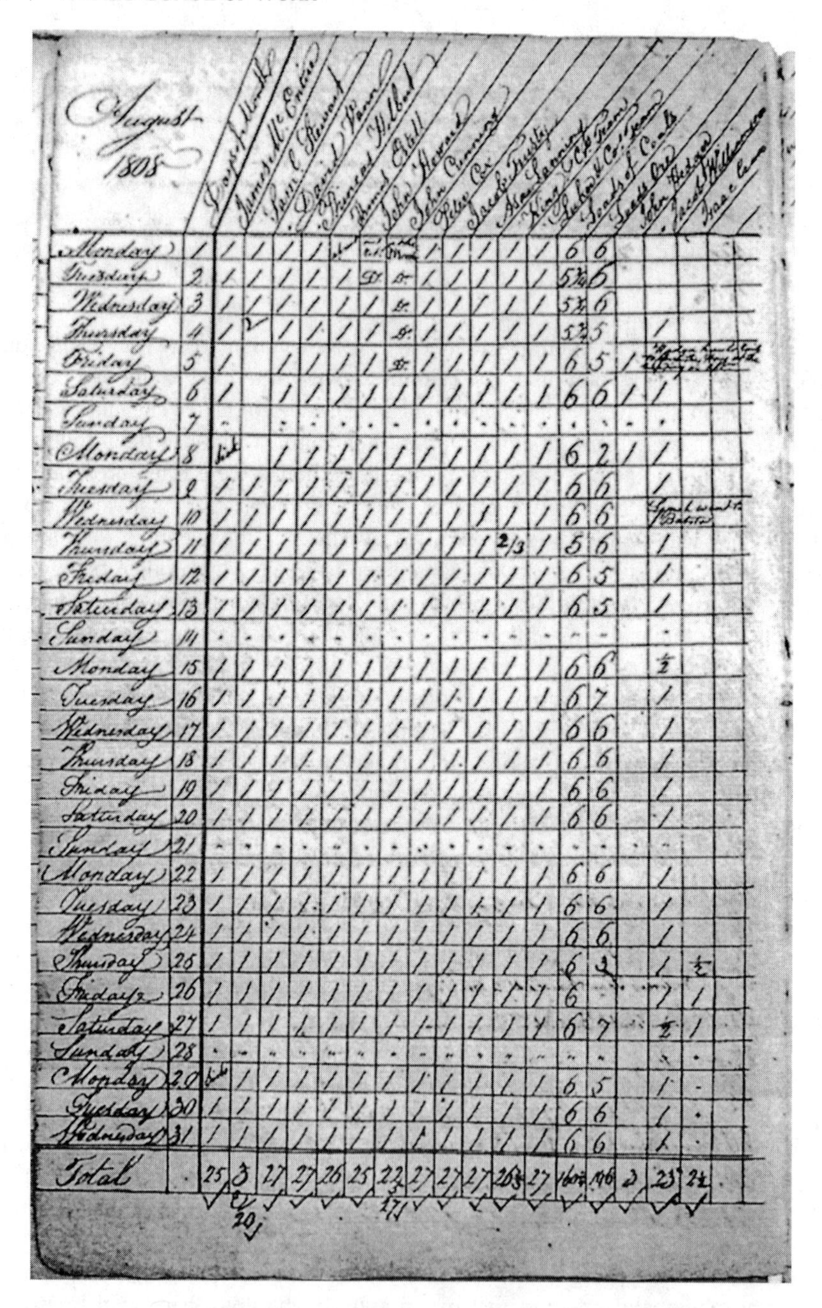

August 1808		Joseph [illegible]	James McEntire	Saml C. Steward	David Vance	Duncan Wilbert	Bernard Estell	John Steward	John Cummins	Peter Cox	Jacob Trusty	Asa Lamont	King & C. Farm	Luke & Co. Teams	Leads of Coals	Leads Ore	John Hedges	Jacob Williamson	Isaac Ca[illegible]
Monday	1	/	/	/	/	[illegible]	T.h.	On	/	/	/	/	/	6	6				
Tuesday	2	/	/	/	/	/	D.	D.	/	/	/	/	/	5¾	6				
Wednesday	3	/	/	/	/	/	/	D.	/	/	/	/	/	5¾	6				
Thursday	4	/	2	/	/	/	/	D.	/	/	/	/	/	5¾	5		/		
Friday	5	/		/	/	/	/	D.	/	/	/	/	/	6	5	1	[illegible]		
Saturday	6	/		/	/	/	/	/	/	/	/	/	/	6	6	/	/		
Sunday	7	-		.	.	.	.	.	.	.	.	.	.	.	.	-	.		
Monday	8	Sick		/	/	/	/	/	/	/	/	/	/	6	2	/	/		
Tuesday	9	/	/	/	/	/	/	/	/	/	/	/	/	6	6		/		
Wednesday	10	/	/	/	/	/	/	/	/	/	/	/	/	6	6		[illegible] went to Balto		
Thursday	11	/	/	/	/	/	/	/	/	/	/	2/3	/	5	6		/		
Friday	12	/	/	/	/	/	/	/	/	/	/	/	/	6	5		/		
Saturday	13	/	/	/	/	/	/	/	/	/	/	/	/	6	5		/		
Sunday	14	.	.	.	.	.	.	.	.	.	.	.	.	-	-		.		
Monday	15	/	/	/	/	/	/	/	/	/	/	/	/	6	6		½		
Tuesday	16	/	/	/	/	/	/	/	/	/	/	/	/	6	7		/		
Wednesday	17	/	/	/	/	/	/	/	/	/	/	/	/	6	6		/		
Thursday	18	/	/	/	/	/	/	/	/	/	/	/	/	6	6		/		
Friday	19	/	/	/	/	/	/	/	/	/	/	/	/	6	6		/		
Saturday	20	/	/	/	/	/	/	/	/	/	/	/	/	6	6		/		
Sunday	21	.	.	.	.	.	.	.	.	.	.	.	.	.	.	.	-		
Monday	22	/	/	/	/	/	/	/	/	/	/	/	/	6	6		/		
Tuesday	23	/	/	/	/	/	/	/	/	/	/	/	/	6	6		/		
Wednesday	24	/	/	/	/	/	/	/	/	/	/	/	/	6	6		/		
Thursday	25	/	/	/	/	/	/	/	/	/	/	/	/	6	3		/	½	
Friday	26	/	/	/	/	/	/	/	/	/	/	/	/	6		.	/	/	
Saturday	27	/	/	/	/	/	/	/	/	/	/	/	/	6	7		½	/	
Sunday	28	.	.	.	.	.	.	.	.	.	.	.	.	.	.		.	.	
Monday	29	Sick	/	/	/	/	/	/	/	/	/	/	/	6	5		/	-	
Tuesday	30	/	/	/	/	/	/	/	/	/	/	/	/	6	6		/	.	
Wednesday	31	/	/	/	/	/	/	/	/	/	/	/	/	6	6		/	.	
Total	.	25	3	27	27	26	25	22	27	27	27	26⅔	27	160½	196	2	23	2½	.

Figure 1. **Martha Furnace Journal, Left Page (1808).** This page listed the names of the workers across the top and the days of the month down the side, recording the employees overall time.

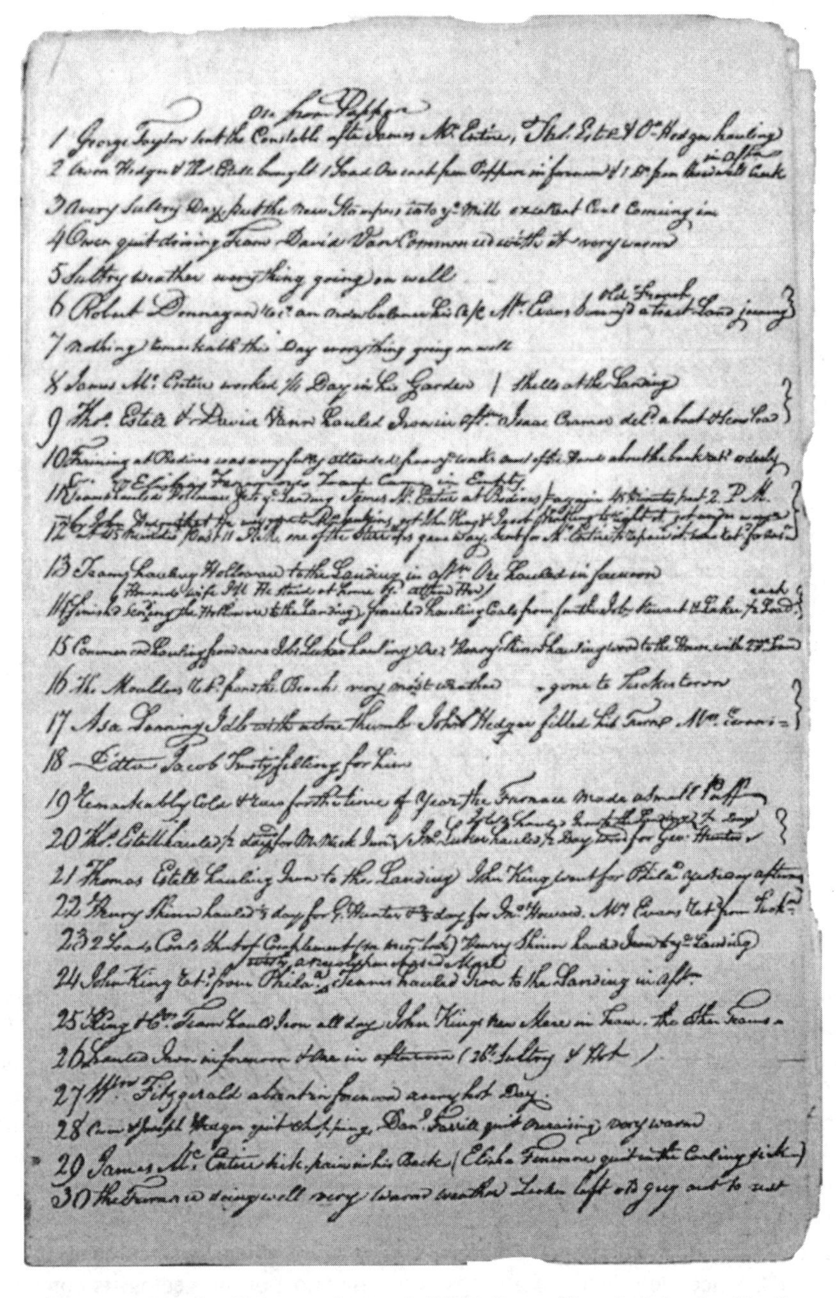

1 George Taylor [illegible] the Constable after James McCartie, Thos. Estell & [illegible] Hedger hauling
2 [illegible] Hedger & Thos. Estell brought 1 load Ore each from [illegible] in forenoon & 1 [illegible] in afternoon
3 A very Sultry Day [illegible] the new [illegible] into ye Mill excellent Coal Coming in
4 [illegible] Team [illegible] very warm
5 Sultry weather everything going on well
6 Robert [illegible]
7 nothing remarkable this Day everything going on well
8 James McCartie worked 1/2 Day in his Garden | [illegible] at the Landing
9 Thos. Estell & David [illegible] hauled Iron in aftn. [illegible]
10 [illegible]
11 [illegible] at the Landing James McCartie [illegible] 2 P.M.
12 [illegible]
13 Team hauling Hollowware to the Landing in aftn. Ore hauled in forenoon
14 [illegible] the Hollowware to the Landing [illegible] hauling Coals [illegible]
15 [illegible]
16 The Moulders [illegible] very moist weather [illegible] gone to Tuckertown
17 Asa [illegible] John Hedger [illegible]
18 Ditto Jacob [illegible] filling for him
19 Remarkably cold & raw for the time of Year the Furnace made a small Puff
20 Thos. Estell hauled 1/2 day for [illegible] hauled 1/2 Day [illegible] for Geo. Hunter
21 Thomas Estell hauling Iron to the Landing John King went for Phila. yesterday afternoon
22 Henry [illegible] hauled 1/2 day for G. Hunter & 1/2 day for Jno. Howard. Mr. Evans [illegible]
23 2 loads Coal [illegible] hauled Iron to ye Landing
24 John King [illegible] from Phila. Teams hauled Iron to the Landing in aft.
25 King & [illegible] Team hauled Iron all day John King's new Mare [illegible] the other Teams
26 Hauled Iron in forenoon [illegible] in afternoon (26th Sultry & Hot)
27 Wm. Fitzgerald absent in forenoon a very hot Day
28 [illegible] Hedger quit [illegible] very warm
29 James McCartie [illegible] pain in his Back [illegible]
30 The Furnace is doing well very warm weather [illegible]

Figure 2. **Martha Furnace Journal, Right Page.** The clerk for the Martha Furnace kept a daily account of significant activity in the ironworks, thereby capturing a portrait of life in a nineteenth century ironmaking village.

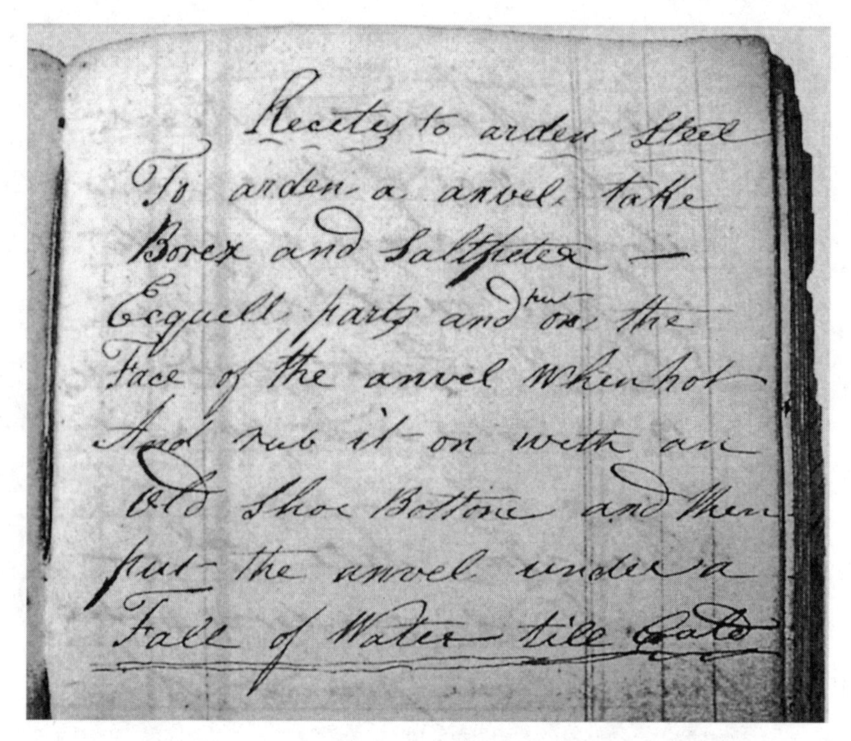

Recites to arden Steel
To arden a anvel take
Borex and Saltpeter —
Ecquell parts and put on the
Face of the anvel when hot
And rub it on with an
Old Shoe Bottome and then
put the anvel under a
Fall of Water till Cold

Figure 3. **George Nock—His Book, Ramapo Works (1837).** This small journal began by recording transactions but changed to recording recipes for nineteenth century metallurgical processes.

STATE REPORTS

In the mid-nineteenth century there was an explosion of publications about the iron industry. States began publishing geological reports about their natural resources, commercial publishers printed books for a reading public, and weekly journals carried current news to those in the business. The state reports were the first publications to be potentially useful to ironmakers. "As educated individuals turned their attention to the application of science to economic development in the early 19th century, some of them convinced their state legislatures to appropriate money for geological surveys intended to find economically useful mineral resources" [4, p. 31]. The early state reports were meant to be persuasive rhetorically in order to encourage the exploitation of the mineral deposits, and in some cases they succeeded. Since the science of geology was new, most of the geologists commissioned to write the reports were more interested in furthering their knowledge of Earth's structure than helping industry to exploit its resources. Consequently, the funding for these early state reports was sporadic and often withdrawn.

The first extensive state geological report was by J. H. Alexander, the topographical engineer of Maryland. His report, published in 1840, was quickly followed with multivolume state reports in Massachusetts and New York. Many of these geological reports intersected with the field of mining, and thus they provided information about minerals, machinery, and topography that was valuable to the growing iron industry. The one that contained the most information about the iron industry was William Kitchell's *Second Annual Report on the Geological Survey of the State of New Jersey for the Year 1855*. Kitchell (1827–1861) focused on the mineral industry. He located and documented iron mines and ironworks, describing processes and machinery that had not yet been captured in writing. He also hired John Hermann Carmiencke (1810–1867), a Hudson River School painter, to accompany the surveyors and produce images of the mines and the miners themselves. Carmiencke sketched in the field and then transferred the sketches to wood, which were then cut by a professional engraver [18]. These images are important because they are early images of an otherwise undocumented industry (Figure 4).

Commercial Books

In the 1850s, even though technical writing was still not used in the workplace, commercial publishers found that people would pay for technical knowledge. They started printing books on technical subjects for general audiences. Like Agricola's and Biringuccio's 16th century volumes, these books were richly illustrated. However, the 19th century volumes were less expensive, contained more detailed illustrations and reached a wider audience. The additional detail in the illustrations came from improvements in wood engraving techniques (Figures 5 and 6). The improvements were the use of an engraver's burin (instead of a knife) and the subsequent ability to cut harder wood against the grain that could withstand longer print runs [19]. These detailed illustrations hastened the pace of knowledge transfer across broad segments of the population. They are part of the shift from prediscursive, physical and verbal technical communication to using writing and drawing as essential tools in the workplace.

Henry Carey Baird & Co., Industrial Publishers, started publishing a series of technical and industrial manuals in 1849, which eventually included hundreds of titles and lasted into the twentieth century. Tebeaux notes that during the Renaissance, "printers soon saw that a market existed for instructional books that covered a variety of practical subjects" [15, p. 10]. A similar reading public emerged in nineteenth-century America. The earliest volumes about the iron industry were written by a German immigrant, Frederick Overman. His *The Manufacture of Iron* was the most widely distributed of his volumes and can still be found in libraries today (Figure 7). Overman published two additional books on metallurgy in the next two years, and many other authors wrote about other

Figure 4. **Woodcut of Drawing by Herman Carmiencke in William Kitchell's Second Annual Report (1856).** Carmienckie, a Hudson River School painter, turned his artistic eye to the early American iron industry and its machinery.

aspects of the iron industry, such as metal working, metallurgy, and molding as well. A reading public had emerged that was hungry for books about technology.

PUBLICATIONS OF PROFESSIONAL ASSOCIATIONS

Writing, as a way of thinking and communicating, was central in the formation of professional associations that studied, discussed, and attempted to solve problems for the new industries. The purposes of the associations, based on the scientific societies in Europe, were "to promote the Arts and Sciences connected with the economical production of the useful minerals and metals, and the welfare of those employed in these industries, by means of meetings for social

Figure 74. Goldbeaters at work on a duplex plate of gold and silver.

Figure 5. **A Woodcut from Biringuccio's *Pirotechnia* (1540).** The woodcut was made with a knife along the grain of soft wood and thus the detail is simplified.

intercourse, and the reading and discussion of professional papers, and to circulate, by means of publications among its members and associates, the information thus obtained" [20, p. xvii]. The writing, reading, and publication of papers was a major method of knowledge exchange on a broad scale. Meeting where geographically distant ironmakers could meet, tour factories, and exchange information face-to-face.

The American Iron Association (later the American Iron and Steel Association; now called the American Iron and Steel Institute) held its first national meeting in 1849. They were shortly followed by the American Institute of Mining Engineers (AIME) in 1870, the United States Association of Charcoal Iron Workers in 1880, the American Society of Mechanical Engineers (ASME) in 1880, and the American Society for Testing Materials (ASTM) in 1898. These are only a fraction of the professional organizations that formed in the nineteenth century: with railroad transportation it was possible, for the first time, to hold meetings with a geographically disparate group on a regular basis. The associations published newspapers, bulletins, directories and transactions (or proceedings) that recorded meetings, paper presentations, dinners and trips to industrial sites, as well as transcriptions of the discussions that followed the reading of the papers (Figure 8). The formation of these groups was roughly concurrent with the beginnings of record keeping on the factory floor at Lukens Steel. Literacy was already a necessity for the members of the professional associations as they traded ideas, and gradually literacy was becoming useful on the factory floor as well. The importance of the professional associations, however, was that

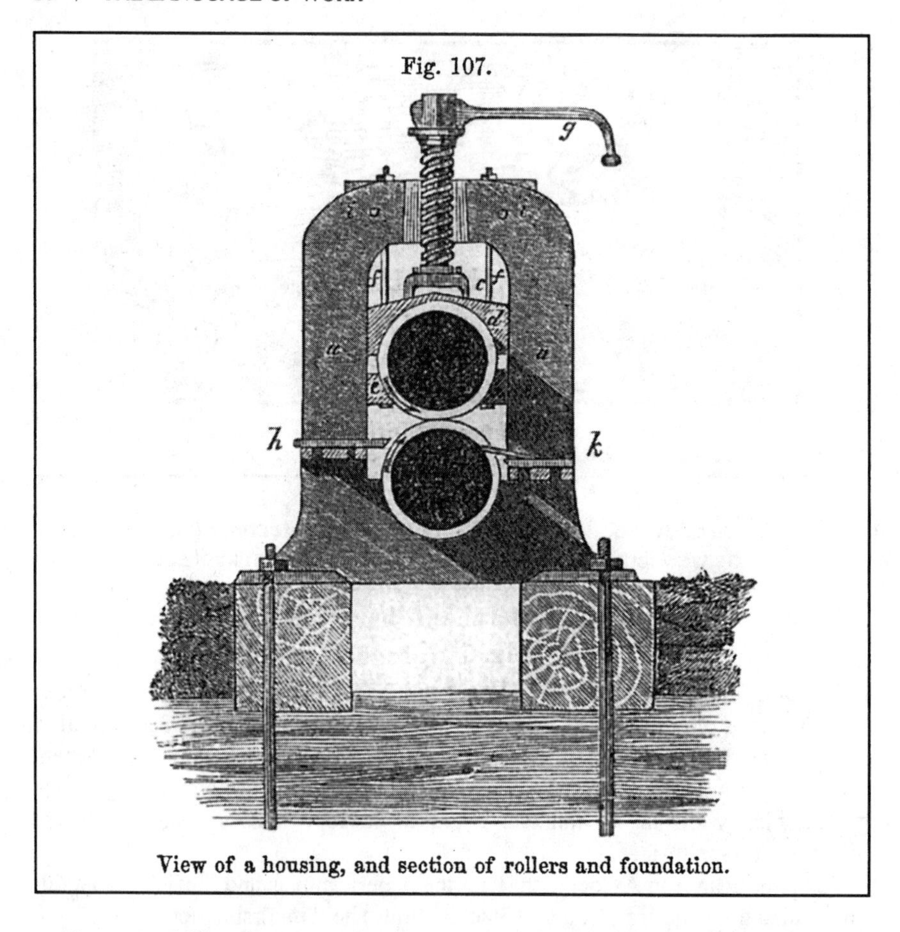

Fig. 107.
g
h
k
View of a housing, and section of rollers and foundation.

Figure 6. **A Woodcut from Overman's Manufacture (1854).** This woodcut was made with an etching burin against the end grain of wood (such as can be seen in the exposed wooden housing above) creating a more detailed image that could withstand longer print runs.

they hastened the rate of knowledge transfer. Their publications—directories, statistics, bulletins, papers, and proceedings—contributed to the rapid development and improvement of early steel technology. Writing had begun to permeate the working world.

ILLUSTRATED TRADE NEWSPAPERS

In the mid-nineteenth century illustrated newspapers appeared on a variety of professional topics. The first to specialize in a single industry was the *American*

THE

MANUFACTURE OF IRON,

IN ALL ITS VARIOUS BRANCHES.

INCLUDING

A DESCRIPTION OF WOOD-CUTTING, COAL-DIGGING, AND THE BURNING OF CHARCOAL AND COKE; THE DIGGING AND ROASTING OF IRON ORE; THE BUILDING AND MANAGEMENT OF BLAST FURNACES, WORKING BY CHARCOAL, COKE, OR ANTHRACITE; THE REFINING OF IRON, AND THE CONVERSION OF THE CRUDE INTO WROUGHT IRON BY CHARCOAL FORGES AND PUDDLING FURNACES.

ALSO

A DESCRIPTION OF FORGE HAMMERS, ROLLING MILLS, BLAST MACHINES, HOT BLAST, ETC. ETC.

TO WHICH IS ADDED

AN ESSAY ON THE MANUFACTURE OF STEEL.

BY FREDERICK OVERMAN,

MINING ENGINEER.

Phoenixville Iron Works.

WITH ONE HUNDRED AND FIFTY WOOD ENGRAVINGS.

Third Edition, Revised.

Figure 7. **Frederick Overman's *Manufacture of Iron* (1854).** Overman was the first to publish general commercial books about the iron industry in America. This volume was very popular and went through many editions.

CONTENTS.

PAGE

Officers, Honorary Members and "Foreign" Members, ... vii
Members and Associates, ... x
Past and Present Officers, ... lxxiv
List of Meetings, ... lxxvi
Publications, ... lxxviii
Rules, ... lxxxi

PROCEEDINGS.

Condensed proceedings of the Chicago meeting, being part of the International Engineering Congress, August, 1893. (For full proceedings, see vol. xxii.) lxxxv

PAPERS.

A New Process for the Production of Pig-Iron, Refined Iron, Ingot-Metal and Weld-Metal, by Alexander Sattman and Anton Homatsch, . . . 3
The Micro-structure of Ingot-Iron in Cast Ingots, by A. Martens (See Discussion, "Physics of Steel," p. 608), ... 37
Experimental Investigations on the "Loss of Head" of Air-Currents in Underground Workings, by D. Murgue, ... 63
Further Observations on the Relations Between the Chemical Constitution and Physical Character of Steel, by William R. Webster (See Discussion, "Physics of Steel," p. 608), ... 113
The Consumption of Fuel in the Taylor Gas-Producer Plants at the Aspen and Marsac Mills Compared, by C. A. Stetefeldt (See Discussion, p. 585), . 134
The Limitations of the Gold Stamp-Mill, by T. A. Rickard (See Discussion, p. 545), ... 137
Iron Alloys with Special Reference to Manganese Steel, by R. H. Hadfield, 148
The Genesis of Ore-Deposits, by F. Pošepný (See Discussion, p. 587), . . . 197
Review of American Blast-Furnace Practice, by E. C. Potter (See Discussion, p. 577), ... 370
Sulphur in Cast-Iron, by W. J. Keep, ... 382
Electricity in Mining, by F. O. Blackwell, ... 399
Recent Advances in Pyrometry, by W. C. Roberts-Austin, ... 407
The Growth of American Mining-Schools and their Relation to the Mining Industry, by Samuel B. Christy (See Discussion, p. 657), ... 444
The Heat-Treatment of Steel, by Henry M. Howe (See Discussion, "Physics of Steel," p. 608), ... 466

DISCUSSIONS.

Discussion of paper of Mr. Rickard (See p. 137), ... 545
Discussion of paper of Mr. Potter (See p. 370), ... 577
Discussion of paper of Mr. Stetefeldt (See p. 134), ... 585
Discussion of paper of Prof. Posepuy (See p. 197), ... 587
Discussion: The Physics of Steel, ... 608
Discussion of paper of Prof. Christy (See p. 444), ... 657

Figure 8. **Sample of the Table of Contents for the *Transactions of the American Institute of Mining Engineers* (1893).** The first meeting of the AIME was in 1871 and the first Transactions were published in 1873, recording meetings, papers, and discussions.

Railroad Journal, started in 1832 [21, p. 1]. However, the journals were flexible and often broadened or narrowed their focus to target a specific readership. In 1849 the *American Railroad Journal* changed its name to the *American Railroad Journal and Iron Manufacturers and Mining Gazette* and added material about the iron industry. In 1867 the *Iron Trade Review* began publication. It changed its name to *Steel* in 1930, when iron was no longer a major product, and changed it again to *Industry Week* when the steel industry in America went into decline. There was also *Iron, Iron Age, Iron Era*, and others.

One of the earliest journals about the iron industry was the *Engineering and Mining Journal* (EMJ). Originally it was the *American Journal of Mining, Milling, Oil-Boring, Geology, Mineralogy, Metallurgy, etc.* (Figure 9). At first it contained articles about the deposits of gold and silver in the American west. In 1868 Rossiter W. Raymond, the eventual secretary of AIME, took over the editorship and the title changed to the *Engineering and Mining Journal.* The weekly published information about a variety of topics, including reports on mining activities from every region of the country (which arrived by steamship); notes on inventions; patents pending; discussions of geology, machinery, and chemistry; prices for coal, securities, gold, and other minerals; and advertisements. After Raymond took over the editorship, EMJ became closely tied to AIME. It, too, published the papers read at the meetings and discussions that followed them, thus ensuring the maximum amount of exposure. In one issue of EMJ, the editor wrote, "It has long been the opinion of those familiar with the transactions of the Institute that the discussions which follow the reading of papers are frequently more valuable than the papers themselves" [22, p. 195]. Tacit knowledge, embodied by experts in the workplace, was in the process of being transferred to codified knowledge, written and drawn in a specific form. The transcribed discussions are a perfect example of a bridge from spoken to written knowledge exchange.

EARLY TWENTIETH-CENTURY TECHNICAL COMMUNICATION

The types of technical communication publications that emerged in the nineteenth century were joined in the twentieth century by many others. Businesses started publishing product guides, which functioned both as informational manuals and as advertising tools. Industry started using requests for proposals (RFPs) and proposals, consultant reports, advertising, and many other forms, some of which will be discussed later. These publications laid the foundation for a literate work environment; by the twentieth century, technical communication had permeated the culture as a whole. It was no longer the domain of professional engineers and business owners; a general literacy, such as being able to understand lists as well as create and interpret drawings, was required by an increasing number of workers; and advanced literacy, such as the ability to read,

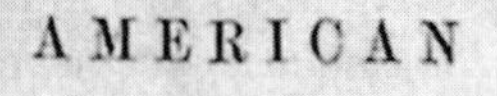

AMERICAN

JOURNAL OF MINING,

Milling, Oil Boring, Geology, Mineralogy, Metallurgy, etc.

GEORGE FRANCIS DAWSON, EDITOR.

PARK BUILDING.

VOLUME II.—OCTOBER, 1866, TO MARCH, 1867.

NEW YORK:

WESTERN & COMPANY, PUBLISHERS AND PROPRIETORS.

37 PARK ROW AND 145 NASSAU STREET.

Figure 9. **Early Version of the Engineering and Mining Journal (1866).** The cover of this large-format weekly newspaper shows the journal headquarters.

LAYING OUT FOR WELDED CONSTRUCTION 279

along the outside edge of outside tube holes. Edge of plate must be bevelled 30-degrees. New plate to be applied as shown with sufficient clamps to prevent warping and buckling; seams should be tack welded, tacks 3 inches wide about every 12 inches apart as indicated by arrow points and numbers on Fig. 490, 1 to 14. When sheet has been securely tacked, proceed to weld as follows: Weld space between tacks 1 and 5, then weld space between 4 and 8, next between 3 and 7, then between 2 and 6. Then between 1 and 3, then 2 and 4. Then between 10 and 14, next between 12 and 13. Then between 9 and 5, next between 11 and 7. Then weld all other spaces until seam is complete.

The above system of alternating the welding to opposite points insures that plate will remain in position and correct opening preserved for welding. The weld should be reinforced about 20% on the water side.

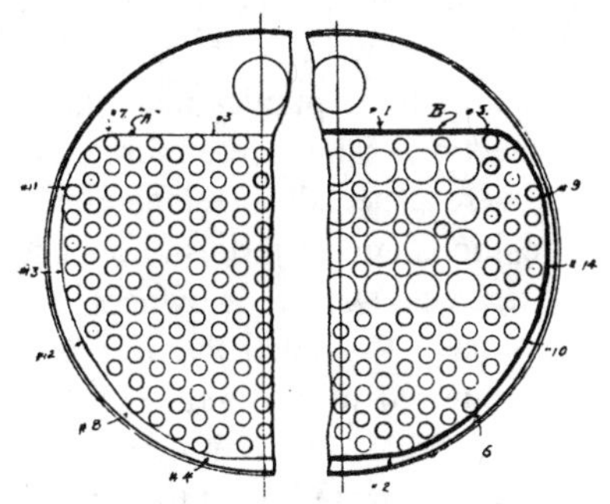

Fig. 490.—Method of Renewing Front Tube Sheet Area

Fig. 491 illustrates another method of applying a new front tube sheet when the flange on the old tube sheet is still good. The old tube sheet is cut above the flange and a new sheet is set in and welded around circumference of the flange as illustrated in Fig. 491. 5/8 inch gusset plates are set in and welded to the front tube sheet and the flange at intervals of 6 inches to 8 inches apart.

Thermic Syphons. Any method of welding syphons to the crown sheet that will properly provide for expansion and prevent creeping as the welding progresses, is satisfactory.

The following backstep method as illustrated in Fig. 492, is suggested to be used with either the electric or acetylene processes of welding butt joints. Plates should be properly clamped and braced to prevent undue movement and distortion. The ends should then be welded; beginning at No. 1 on the diagram, coming around the corner from 6 inches to 8 inches. Follow with No. 2, then 3 and 4 at the front end, then back to 5, etc., using

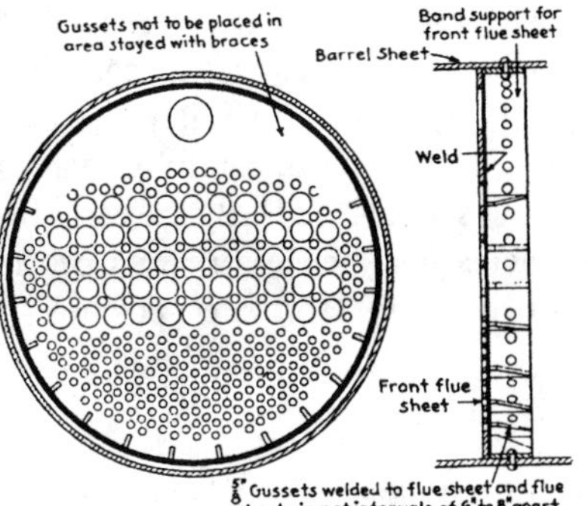

Fig. 491.—Straight Welded Front Tube Sheet

the "Back-Step method. Going on equal distance on either seam to distribute the stress. The length of "steps" depends on the rigidity of sheets. Where radial staybolts have been applied in crown sheet the "steps" should be from 12 inches to 14 inches long, and where radial staybolts have not been applied, the "steps" may be longer.

The "Back-Step" method of welding may also be used for welding the diaphragm plate to the Tube Sheet.

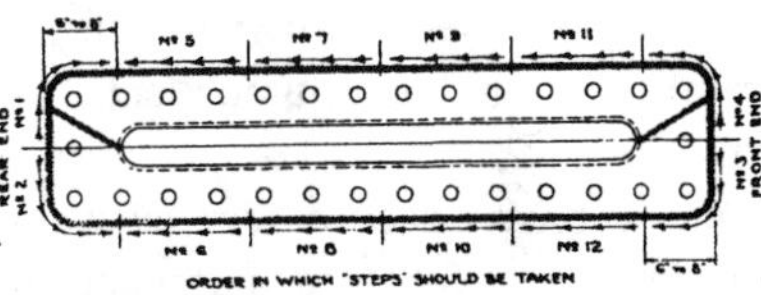

Fig. 492.—Method of Welding Syphon in Crown Sheet

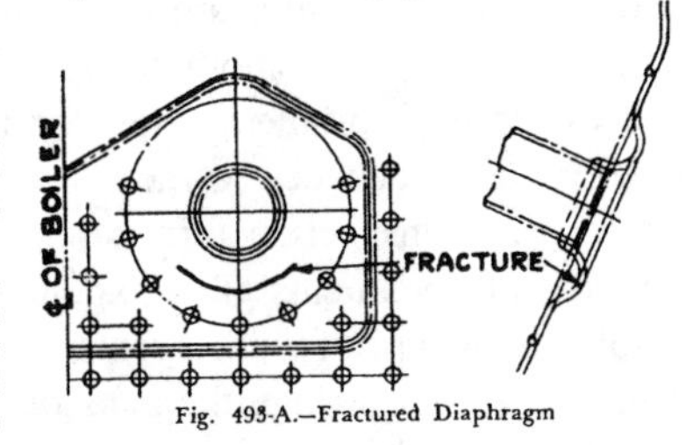

Fig. 493-A.—Fractured Diaphragm

Figure 10. **Page from *Laying Out for Boiler Makers and Plate Fabricators.*** This is an example of a highly-specific technical manual containing both text and drawings [24].

interpret, and write, led to management positions. Colleges began forming and advertising their mining and metallurgy programs in the 1860s, and new commercial publishers emerged which produced educational texts. For instance, the Simmons-Boardman Publishing Company first published *Laying Out for Boiler Makers and Plate Fabricators* in 1907 [23], a book that illustrates the end-use for some of Lukens' product. The book contains both text and pictures so that it would be accessible to both literate and semiliterate workers at different levels (Figure 10). The Wiley Technical Series started at about the same time to provide "A series of carefully adapted texts for use in technical, vocational and industrial schools" [24]. By the early twentieth century, technical communication was increasingly necessary to workers. Literacy had become a skill that led to more powerful (and safer) positions.

* * * * *

The preceding is the background against which the emergence of technical communication at Lukens Steel took place. The owners of the firm used many of the resources listed above—they read the trade publications, published articles in journals, and presented papers at meetings. They were more educated than the majority of ironmasters because they were a specialty steel manufacturer, dependent on using specific chemical compositions and materials knowledge in their manufacturing process. Their advanced degrees (two of the owners, Charles Lukens and Charles Huston, were medical doctors) enabled them to produce a more complex product and that, in turn, required more literacy from their employees. At first, the owners encouraged the foremen to visit other plants to gain information, but as time went on they also encouraged them to subscribe to journals and attend trade-association meetings. At the end of the nineteenth century, intensive record keeping was required of foremen, but by the twentieth century workers, foremen, and management used writing and drawing to describe and solve problems, and share knowledge. More voices had joined the technological conversation and more minds used writing and drawing to come up with new ideas and solve problems. New forms of technical communication emerged—handwritten notes between workers, management, and owners; graphs; blueprints; test reports and inspection forms; scientific analysis and correspondence—and the amount of documents grew exponentially. In 1915, when carbon paper was available, a new type of worker joined the plant, the stenographer typist. This new worker was able to bridge the gaps between levels of literacy and turn spoken language into readable documents of great length and detail. The stenographer typist, later called the secretary, embedded literacy within the organization.

REFERENCES

1. J. G. Landels, Review: *A Mine of Information, Mining and Metallurgy in the Greek and Roman World* by J. F. Healy, *The Classical Review, 29*:2, pp. 297-300, 1979.

2. A. Y. al-Hassan and D. R. Hill, *Islamic Technology: An Illustrated History*, Cambridge University Press, New York, 1986.
3. J. Yates, *Control Through Communication: The Rise of System in American Management*, Johns Hopkins University Press, Baltimore, Maryland, 1989.
4. R. B. Gordon, *American Iron: 1607-1900*, Johns Hopkins University Press, Baltimore, Maryland, 1996.
5. P. O. Long, *Openness, Secrecy, Authorship: Technical Arts and the Culture of Knowledge from Antiquity to the Renaissance*, Johns Hopkins University Press, Baltimore, Maryland, 2001.
6. A. Wilson, Machines, Power and the Ancient Economy, *The Journal of Roman Studies, 92,* pp. 1-32, 2002.
7. G. Agricola, *De Re Metallica*, Dover, New York, 1950.
8. W. Byrd, *The Westover Manuscripts: Containing the History of the Dividing Line Betwixt Virginia and North Carolina: A Journey to the Land of Eden, A.D. 1733; A Progress to the Mines. Written from 1728 to 1736, and Now First Published*, accessed June 19, 2007 from http://docsouth.unc.edu/nc/byrd/menu.html
9. A. Spotswood, *Iron Works at Tuball: Terms and Conditions for Their Lease*, 1945, L. J. Cappon (ed.), University of Virginia, Charlottesville, Virginia, 1739.
10. S. G. Hermelin, *Report About the Mines in the United States of America, 1783*, 1931, A. Johnson (trans. and ed.), John Morton Memorial Museum, Philadelphia, Pennsylvania, 1783.
11. P. Hasenclever, *The Remarkable Case of Peter Hasenclever, Merchant,* 1970, North Jersey Highlands Historical Society, New Jersey, 1773.
12. R. Erskine, "Shrewsbury, October 15, 1770," *Manor and Forges of Ringwood Compilation, 1759-1923*, New Jersey State Archives, Trenton, New Jersey, 1770.
13. C. S. Johnson, Prediscursive Technical Communication in the Early American Iron Industry, *Technical Communication Quarterly, 15*:2, pp. 171-189, 2006.
14. C. B. Dew, *Bond of Iron: Master and Slave at Buffalo Forge*, W. W. Norton & Co., New York, 1994.
15. E. Tebeaux, Visual Texts: Format and the Evolution of English Accounting Texts, 1100-1700, *Journal of Technical Writing and Communication, 30*:4, pp. 307-341, 2000.
16. H. H. Bisbee and R. B. Cloesar, *Martha, 1808-1815: The Complete Martha Furnace Diary and Journal*, Henry H. Bisbee, Burlington, Vermont, 1976.
17. G. Nock, *George Nock: His Book Ramapo Works*, Hagley Museum and Library, Wilmington, Delaware, 1837.
18. W. Kitchell, *Second Annual Report on the Geological Survey of the State of New Jersey for the Year 1855,* The True American Office, Trenton, New Jersey, 1856.
19. M. Twyman, *The British Library Guide to Printing History and Techniques*, University of Toronto Press, Toronto, Canada, 1998.
20. *Transactions of the American Institute of Mining Engineers*, The Institute, Philadelphia, Pennsylvania, 1873.
21. K. L. Endres, *The Roots of the Trade Press: The American Railroad Journal and the Professionalization of an Industry, 1832 to 1840*, Association of Journalism and Mass Communication Convention Presentation, Boston, Massachusetts, 1991.

22. R. W. Rossiter (ed.), *Engineering and Mining Journal, 12*:13, p. 195, 1871.
23. G. M. Davies, *Laying Out for Boiler Makers and Plate Fabricators* (5th Edition), Simmons-Boardman Publishing Corporation, New York, 1944.
24. F. W. Turner and D. G. Town, *Pattern-Making*, John Wiley & Sons, New York, 1914.

CHAPTER 2

The History of Lukens Steel (1810-1925)

In the history of Lukens Steel we can see the rough edges of the changes in communication over time. During the early and mid-nineteenth century, when industry was young, only the business owners and later, a few employees used writing as a method of communication. The tools were primitive—the quill pen, paper, and later letterbooks—and thus communication was kept at a minimum. This state of affairs lasted until about 1870, when they built more sophisticated technological mills and needed to keep records for inspection and tracking defects, at which point record keeping joined the correspondence and financial journals. It wasn't until 1890, when they built two open-hearth furnaces and the largest rolling mill in the country, that the need for technical communication increased. Consequently, more and more voices from across the plant joined in the social discourse community. They were beginning to rely on writing and drawing to communicate and solve complex problems. This reliance on multiple voices exploded in the early twentieth century to the massive and multiple networks of technical communication that we are familiar with today.

As a single instance of a particular industry, it should be stated that Lukens Steel is unusual. First, since it remained under direct family control for 188 years, the owners were able to save many of their papers over a long period. This situation is not replicated in many other companies that frequently discarded old documents. The Lukens documents provide a microview of the development of technical communication as well as the technology during this time. Second, Lukens Steel was a family-owned business and it was run more like a family than a corporation—their Quaker background served them in their attempt to treat their employees fairly. The result was that multiple families worked at the plant for generations. Lukens was unique in its business philosophy as well. In the beginning, their policies were very conservative: "The partners never changed anything unless the pressure to do so was enormous" [1, p. 177]. However, when complex new technologies increased demand for their product, they met the demand, experimented, and built cutting-edge facilities. Their attitude toward technical communication and its technologies parallels this mixture of conservatism and progressivism: although they introduced the typewriter in 1885, they maintained the cumbersome letterbooks until 1903 and did not hire multiple stenographer typists until 1910. Thus, the dates of their adoption of communication technologies cannot be applied to other industries, which evolved differently.

This chapter provides a brief history of Lukens Steel and its owners, as well as an overview of new plants built between 1810 and 1925. This is background material for understanding how the plant itself and the social organization surrounding it evolved over time. The period includes two wars—the Civil War and World War I—both of which were important to the company's success and expansion. Most importantly for this book, however, the time period spans the shift from craftwork to industrialization. In 1810 ironmaking was still a craft, done by individual experts, with variable results. By the end of the nineteenth century, steel, a specialized form of iron, had largely replaced it. By 1925 steelmaking was a complex set of actions, interactions, and reactions, some of which were understood and some of which were not. Technical communication evolved at Lukens Steel in order to communicate complex problems and propose experiments and solutions. And, as communication within the factory increased, so did the communication with consulting engineers, government inspectors, professional associations and other experts as they worked together to define standards and find the best metallurgical solutions to make a product as important as boiler plate for ship and railroad engines as safe as it could be.

THE EDUCATION OF REBECCA LUKENS AT THE FEDERAL SLITTING MILL

The most unique feature of Lukens Steel is that it was set on a sound financial footing by a woman who, despite being pregnant and with three children, took over the business when her husband died in 1825. For over 22 years Rebecca Lukens negotiated the tempestuous financial markets of the nineteenth century iron industry and set the business on a course of "riskless sufficiency" that enabled them to survive [1, p. iv]. The financial foundation that she created was followed by the continued success of four subsequent generations. Family and work were one to Rebecca Lukens [2, p. 277]. Family and work were one to the generations that followed as well.

Lukens Steel began as the Brandywine Iron Works and Nail Factory in 1810, but Rebecca Lukens (1794-1854) began her education in the industry at her father's side in the Federal Slitting Mill, started by her father, Isaac Pennock, in 1793. Prior to the Revolutionary War, the iron industry in America was severely curtailed by the Iron Act of 1750, passed by Parliament, which forbade the making of iron products other than pig or bar iron. The act specifically forbade rolling and slitting mills, and so when the war was won, Pennock named his first mill after the government that made it possible. The Federal Slitting Mill, like most industries in the eighteenth century, required water for power, and thus it was set at the edge of a river. The technology in a slitting mill was simple: using bar iron (pig iron refined by hammering or puddling), the workers heated and rolled the raw iron until it was a sheet. Sometimes they reheated it, folding it over and hammering it for extra strength. These sheets were then made into nail rod, hoops for barrels and wagon wheels, and rods that were sold as bar iron for blacksmiths.

Isaac Pennock started the Federal Slitting Mill in 1793 and Rebecca was born in 1794. She was the eldest of six children and "Isaac taught Rebecca to read, write, calculate and ride horseback. She was his constant companion as he tended to his iron business and properties" [3, pp. 3-4]. Thus, from an early age, work and family were inseparable for Rebecca; she learned to run a rolling mill as a child. When Rebecca was 12, she was sent to boarding school in Westtown and later Wilmington, an environment in which she flourished. Later, while visiting an aunt in Philadelphia, she met a physician, Dr. Charles Lukens, with whom she immediately fell in love. Her love was returned and they were married in 1813. In 1814 Lukens joined his father-in-law in operating the Federal Slitting Mill. He worked there until 1816 when he, Rebecca, and their children moved to Coatesville, Pennsylvania to operate the Brandywine Iron Works and Nail Factory, which had also been started by her father in 1810 (see Figure 1).

BRANDYWINE IRON WORKS AND NAIL FACTORY, 1810-1825

In 1800 the town Coatesville did not exist. In 1810 Moses Coates sold his farm to Isaac Pennock and Jesse Kersey, who developed 40 building lots along the main street. This included a sawmill on the Brandywine River, which became the Brandywine Iron Works and Nail Factory. Coatesville, Pennsylvania was a fortuitous location in several respects. First, it lay along the west branch of the Brandywine River, which provided power for a variety of mills. Second, the first major highway in the United States, the Philadelphia Lancaster Turnpike (later called Lincoln Highway), was built through Coatesville in 1794. The road was "the first important turnpike and the first long-distance broken-stone and gravel surface built in America according to formal plans and specifications" [4]. The turnpike crossed the Brandywine River over a bridge and, Coatesville became a toll stop with a hotel. Small though it was, Coatesville has appeared on maps since 1822.

Isaac Pennock and Jesse Kersey converted the sawmill to an ironworks. In 1816 Pennock bought out his partner and leased the mill and land to his son-in-law, Dr. Charles Lukens. Rebecca and Charles moved their family to Coatesville, occupying a house that had been built by the original landowner in

Figure 1. **Brandywine Ironworks and Nail Factory.** This artistic rendition shows the mill after 1834, when the Philadelphia & Columbia Railroad ran through Coatesville.

the second quarter of the eighteenth century and expanded by the Coates family (Figure 2). There Rebecca had six children, three of whom lived to adulthood.

The technology in an early ironworks was both simple and complex: simple in that the process required few sophisticated tools, complex in that the chemical interactions within the metal could not be seen and thus were often unknown. Still, the Brandywine Iron Works and Nail Factory put out a fairly uniform product. When Dr. Lukens took over, they still rolled nail rod, hoops, and rods for blacksmiths, as Isaac Pennock had done; but soon Dr. Charles Lukens realized that there was an untapped market for boiler plate in America, most of which was imported from England. "A specialized iron, boiler plate, came into demand with the widespread adoption of steam engines for powering everything from small machines to steamboats after 1810" [5, p. 34]. Steamboats were first coming into wide use, steam engines were beginning to power industries, and the expansion of railroads required sheet metal as well. Moreover, the idea of metal-hulled ships was being experimented with in Europe. In 1818 Lukens rolled the first boiler plate in America.

Figure 2. **Brandywine Mansion.** The lower end was built during the second quarter of the eighteenth century. The building still stands, attached to the vacant Lukens Cooperative Store.

Lukens supplied the sheet iron for the first iron-hulled steamboat in America, the *Codorus,* built by John Elgar. Elgar wrote to Lukens on March 31, 1825, requesting the sheet iron. The specifications were:

> 350 feet in Length & 2 feet wide
> 58 " " " " 2" 1 in "
> 30 " " " " 1" 11 in "
>
> 438 feet running measure all rolled exactly to the small guage [sic] or 1/12 of an inch thick except 6 sheets of the 2 feet width rolled to fit the large guage [sic] or 1/8 of an inch thick. The iron to be of the best quality and sound, and particularly clear of buckles or bilges that prevent the sheet from lying flat.
>
> Also send 120 straps or hoops 7 feet long 3 in broad & rolled to fit the large guage [sic] or 1/8 of an inch thick, these dimensions must be strictly attended to. These hoops I want of the best picked iron as the[y] are to be turned to a right angle along the middle from end to end, so that the end shows the form 'L'. If the iron is not sound and very tough it will crack along the corner [6, p. 10].

Elgar finished the letter containing the specifications with a reminder to "return the gauge [sic] with the iron." At this time, standards for measuring (and spelling) were flexible and thus people made their own rules. Also, at this time the only form of testing iron was to notch and break a bar to judge it by the appearance of the fracture [7]. Nevertheless, it was a time of exciting technological possibilities and experimentation. Lukens entered the orders into his book on April 13, and he took payment on May 10. On April 23 Elgar ordered more iron but tragedy struck when Dr. Lukens suddenly died of fever. The early days of the ironworks under his management are recorded in the company journals and incoming handwritten correspondence.

REBECCA LUKENS, IRONMASTER, 1825-1847

Rebecca Lukens (Figure 3) had two young children and a teenage girl when her husband died. She had just gone through two difficult deaths—her toddler Charles and her father Isaac had both died in 1824. She was also pregnant with her last child when her husband died in 1825. She felt, on her husband's death, that she had no choice but to continue the business. Years later she wrote in her personal journal:

> In the summer of 1825, I lost my dear and excellent husband. During the period of our being here the iron business had been very poor . . . in our constant expense in repairing the Works, it was utterly impossible there should be support left for the young and helpless family now dependent solely on me . . . [Dr. Lukens] was sanguine in his hopes for success, and this

Figure 3. **Rebecca Lukens.** This image of Rebecca Lukens, in her later years, is from the family album.
Courtesy Hagley Museum and Library, Acc. 50, Pictorial Collections.

> was his dying request—he wished me to continue and I promised to comply. Indeed I knew well I must do something for the children around me . . . I will not dwell on my feelings, when I began to look around me . . . [but] Necessity is a stern task mistress; and my every want gave me courage; besides . . . where else could I go and live . . . Dr. Lukens had many good and firm friends, and they all stood by me . . . the workmen were tried and faithful, and so with some fear but more courage, I began to struggle for a livelihood . . . now I look back and wonder at my daring [3, pp. 8-9].

Rebecca gradually got the business on a sound financial footing. She enlisted her brother-in-law as her manager and received the necessary materials on credit [1, p. 26]. Charles Brooke, owner of the Hibernia ironworks, upriver from Coatesville, supplied her with bar iron on credit and loaned her money as well [8]. By 1834 she had rebuilt many parts of the mill. She wrote, "the mill had been entirely remodeled, and rebuilt from the very foundation. Dam entirely newly built, Wheels put in, castings, furnaces, mill head, mill house much larger, all were built anew; not a vestige of the old remained . . . I have thoroughly repaired the mansion house, built good and substantial tenant houses for my workmen, and put much lime and fencing on the farm. . . . I had built a very superior mill, though a plain one, and our character for making boiler plate stood first in the market, hence we had as much business as we could do" [3, pp. 9-10]. Rebecca Lukens kept records of the operations of the Brandywine Iron Works in a series of journals and legal documents, but she saved only incoming, not outgoing correspondence.

A major contributor to the success of Lukens Steel was the east-west Philadelphia & Columbia Railroad, which was built through Coatesville in 1834 (Figure 4). "The railroads' capability to provide relatively cheap and rapid transport across a seemingly vast land, to stitch together expanding markets by linking suppliers and users of commodities at places where waterways were unknown and other transport was impractical, was a technological and economic feat of unprecedented importance" [9, p. 67]. Prior to 1834 Lukens had to import coal via the Susquehanna River and canal to Columbia, and then via mule wagon to Coatesville. All of the iron billets, or blooms, and all of their outgoing product was carted by wagon. The railroad enabled Lukens to import coal and bar iron and transport their products with ease.

The mill operated successfully without technological change until 1853, when they added a furnace and enlarged the rolls from 48- to 60-inches, allowing them to roll wider sheets. The mill had one set of rolls, two heating furnaces, shears, and an anvil. The 17 employees were managed by Rebecca's brother-in-law, Solomon Lukens. She provided housing for the workers and invested in real estate as well. She also began a policy of "riskless sufficiency," which was to stop producing product when the demand was low and set the men to doing repairs instead [1, p. iv]. Production would only begin again when the price of bar iron

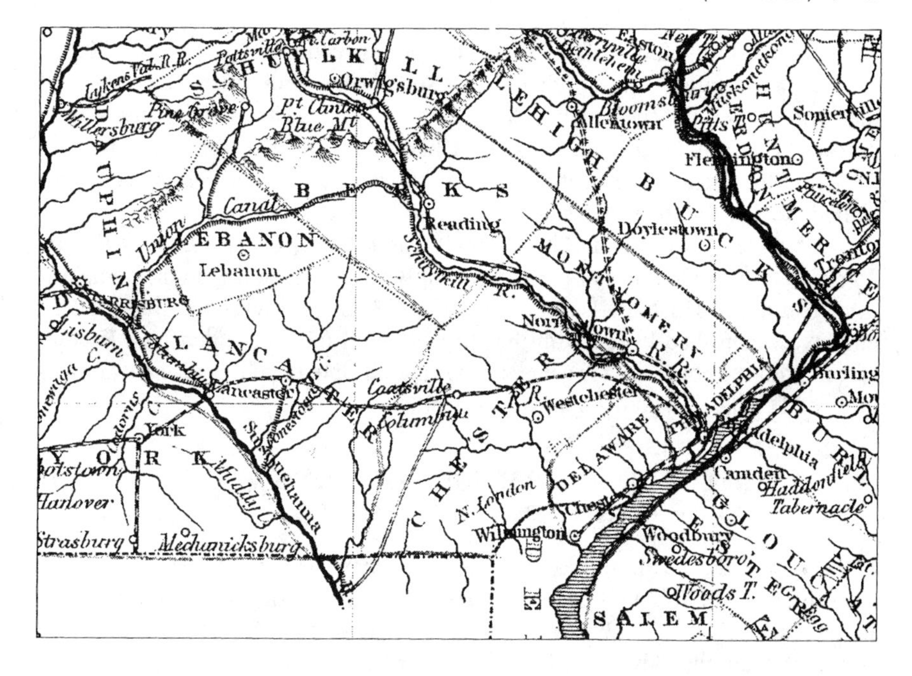

Figure 4. **Map of Pennsylvania (1842).** This map shows the route of the Philadelphia & Columbia Railroad running through Coatesville. Ease of transportation set the stage for continued success.

was acceptable and the product saleable. Thus, the mill that Rebecca built was sufficient to operate with minimal changes for over 30 years.

By the 1850s the Brandywine Iron Works had stopped producing nail rod and hoops and was specializing in boiler plate. They served a national market with agents in Philadelphia, New Orleans, New York, Albany, and Boston. The agents took a 5% commission on the sales. Lukens also sold directly to some companies such as Baldwin Locomotive Works, but her agents were their main source of business and also her source of information about market conditions [1, p. 29]. The amount of correspondence increased at this time as the agents sent specifications back and forth. Rebecca had taken several partners for varying amounts of time, who assisted with the correspondence and financial record keeping. In 1847, outgoing, as well as incoming correspondence was kept in longhand copies. In 1850 they began keeping outgoing correspondence in letterbooks.

In addition to the property Rebecca Lukens rented and the mortgages she held, she "opened a store and a warehouse which combined the functions of a depot and freight agency" [1, p. 32]. At her death, she was the wealthiest woman in Chester County: her estate was valued at $107,952.28, which is worth about $2.7 million today. She left trust funds of $20,000 to her two surviving daughters and $30,000 to her granddaughter. And, although her sons-in-law took over the running of the firm, she left the firm itself to her daughters [2, p. 298]. Isabella and Martha co-owned it until Martha sold Isabella her share. Isabella then became sole owner of the plant. From that point forward Isabella and her descendants controlled the distribution of the shares.

The business stayed under direct family control until Charles Lukens Huston Jr. retired as president in 1974. The last family member working in the company, Charles Lukens Huston III, retired in 1991 [13, p. 5]. Moreover, many of the employees who had begun with the Brandywine Iron Works and Nail Factory stayed with the company for generations, and multiple members of their families grew and changed with the company as well.

THE FIRM IN TRANSITION—GIBBONS, HUSTON, AND PENROSE, 1842-1869

Rebecca Lukens referred to the family members who ran the works as "the firm" [2, p. 278]. Solomon Lukens, Charles' brother, ran the day-to-day activities in the mill until 1837. In 1841 Rebecca's eldest daughter, Martha, married Abraham Gibbons, who joined the firm in 1842. In 1847 Rebecca retired from active participation in the business. In 1848 her daughter Isabella married Dr. Charles Huston (who, like Dr. Lukens, had been trained as a medical doctor) and Dr. Huston joined the firm in 1849. Rebecca died in 1854, and Gibbons withdrew in 1857 to start the Bank of Chester Valley. Isabella purchased Martha's share in the firm and Dr. Huston ran it from 1849 until his death in 1897. One of his cousins, Charles Penrose, joined the firm from 1859 to 1881. Dr. Huston's two sons, Abram Francis and Charles Lukens, took positions with the firm as soon as they graduated from college. All of these different handwritings, including those of clerks and important employees, are recorded in the letterbooks. In 1885 Lukens bought their first typewriter and some of the letters began to be typed, especially those of Charles Lukens Huston, Dr. Huston's son who was the works manager, and whose handwriting was especially hard to read.

Although the plant itself was still called the Brandywine Iron Works, the name of the partnerships changed frequently during these years, and these changing names can be seen in correspondence. When Gibbons joined, the firm name became "R. W. Lukens and A. Gibbons, Jr."; when Rebecca retired it became "A. Gibbons, Jr., & Co."; when Huston joined it became "Gibbons and Huston"; when Gibbons left, it became "Huston & Penrose"; and when Penrose died it became "Chas. Huston & Sons" [10]. The plant's name was separate from the partnership names. When Rebecca Lukens died, the partners changed the name from the Brandywine Iron Works to the Lukens Iron Works in her honor. That name remained until 1890 when the firm incorporated as Lukens Iron & Steel Company. The firm renamed the works the Lukens Steel Company in 1917.

Probably the most important thing about this period is that Lukens reaped enormous profits during the Civil War: output doubled and profits sextupled between 1861 and 1864. This occurred without significant new investment or innovation on the part of the firm. In fact, "By 1869 it was remarkably backward, remaining wed to water power and antique metallurgical techniques" [1, p. v]. Moreover, the number of workers was small: 17 men operated the machinery and made the repairs before the Civil War; during the war, that number increased to only 34: "The firm achieved this dramatic rise in production simply by doubling manpower and running the mill harder" [1, p. 131]. In 1870 the firm finally built a new plant. From this point on, although Lukens Iron Works remained technologically conservative and slow to change, of necessity they abandoned the economic model of "riskless sufficiency." Since their workforce grew with their expanding technological sophistication, technical communication began to increase in types and amount as well.

LUKENS IRON WORKS, 1870-1900: INTRODUCING INDUSTRIALISM

In 1869 the Wilmington and Northern Railroad was built through Coatesville running north and south (Figure 5). This railroad brought coal directly to Lukens Steel: it was "a regional carrier born of the need to move anthracite coal from the Pennsylvania mountains to the cities of Philadelphia and New York" [11]. The location of Coatesville was so good for iron and steel products that there were several such companies in the town: the Triadelphia Iron Works, Worth Brothers, Valley Iron Works, Coatesville Boiler Works, Philadelphia Ranger Boiler, and Viaduct and Welded Steel Shapes [10]. In the nineteenth century these iron mills cooperated with each other by refusing to poach skilled workers and by sharing technological information [1, p. 90]. Later, they cooperated by trading wage information as well [12]. Thus, the technical communication in the immediate vicinity usually occurred by direct personal contact—they sometimes even borrowed equipment and supplies from each other.

When Dr. Charles Huston finally did begin to expand the iron works in 1870, the plant grew rapidly. That year they built an 84-inch steam-powered mill and converted the old mill to furnaces for puddling their own billets (Figure 6). The quality of their rolled iron depended, to a great extent, on the quality of the billets they purchased. Once they had the puddling mill, they gained additional control over the billets' size, shape, and chemical composition. The firm's reputation for high-quality boiler-plate iron continued to grow. In 1857 they had begun offering a warranty on their iron: "Under the terms of the guarantee the mill agreed to replace or take back at the original cost any sheet of iron that failed to flange or that contained a flaw due to improper manufacture." Furthermore, "In 1860 the Corliss Steam Engine Company reported that the average tensile strength of Lukens iron was over sixty thousand pounds per square inch, better than the

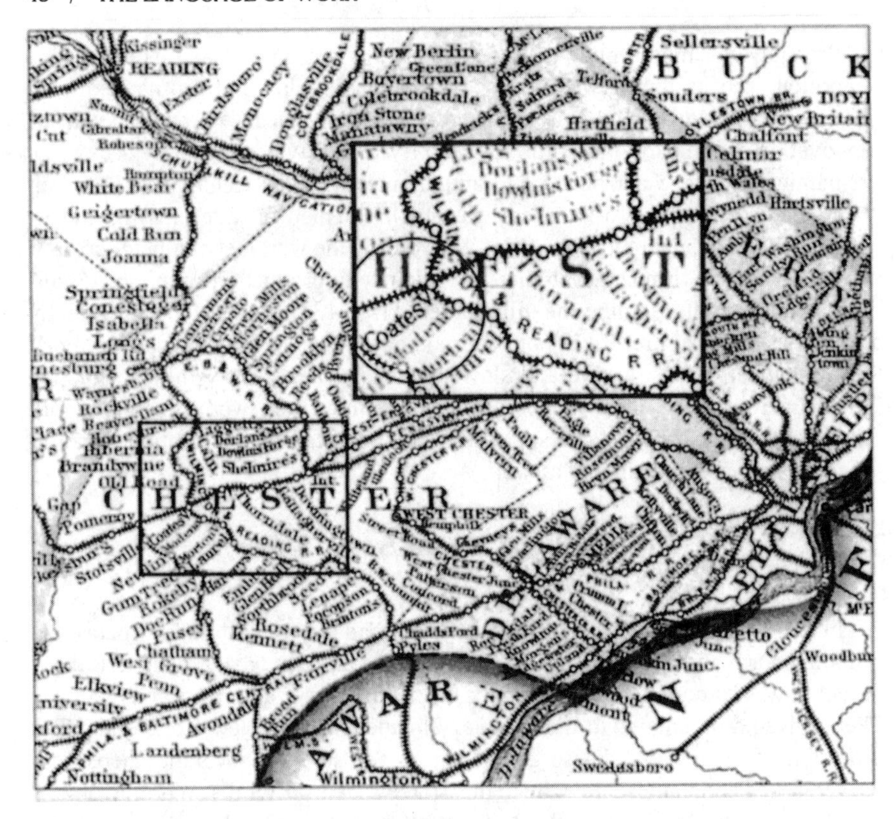

Figure 5. **Gray's Railroad and County Map (1876).** This map shows the location of Coatesville in the midst of multiple railroad networks. Time-to-market decreased considerably and made Coatesville a center of the specialty iron and steel industry.

average of any other lot tested by Corliss up to that time" [1, p. 48]. Their iron production was not perfect; like other iron-plate producers, they had rejected plates returned to them. The difference between Lukens and other rolling mills, however, was that they consistently paid attention to resolving the problems; after 1870 they kept records of the types of defects and the frequency at which they occurred and used those records to trace the source of the problems.

Dr. Huston used writing to communicate the results of his scientific experimentation. His medical training likely led to his scientific approach to the exploration of the nature of iron and steel. In 1875 he acquired a testing machine and took part "in several government-sponsored investigations of boiler explosions

Figure 6. **Lukens Iron Works Company (1878).** The new 84-inch steam-powered rolling mill with two reverberatory furnaces is to the right and the old mill, converted to a puddling mill, is to the left. The office is the two-story building to the right rear.

and became recognized as an authority on the strength and safety of boilers" [13, p. 2]. He published two articles, "The Strength and Ductility of Iron and Steel Boiler Plate at Different Temperatures" and "The Effect of Continued and Progressively Increasing Strain upon Iron" in the *Journal of the Franklin Institute* in 1878 and 1879. He was also involved with the formation of the first specifications of standards for boiler plate. Later, his son Charles Lukens Huston continued his scientific inquiries and continued interacting with governmental, trade, and association representatives in updating the boiler-code safety standards. The testing inspectors, chemists, and other plant employees continued to use the carefully kept data and the scientific tests to trace the source of the defects; it was an ongoing company project.

Lukens survived the economic downturn of 1873, after which the number of employees increased from 34 to 100. Their major markets for boiler plate were the railroads and the shipbuilding industry along the Delaware River. Dr. Huston's eldest son, Abram F. Huston (1852-1930), joined the firm in 1872 and Charles Lukens Huston (1856-1951) joined in 1875 [13, p. 2]. From the beginning the two sons worked at opposite ends of the business: Abram presided over the office and sales divisions, and Charles ran the manufacturing end. Abram Francis Huston was the president of the firm and Charles Lukens Huston was the vice president (Figure 7). Since the majority of the correspondence that was deposited with the Hagley Museum comes from the files of Charles Lukens Huston, he is the pivotal figure in the majority of this analysis.

In 1880 Lukens took another step forward and began rolling steel plate. Initially they ran into trouble, because the steel cracked the rolls that they had previously used for iron. Thus the 84-inch mill was expanded to a two-high roughing and finishing mill with a separate three-high roll train: a two-high mill has rolls that move in opposite directions so that the motion helps to pull the iron or steel through; a three-high mill has three rolls so that the sheet can pass through the rolls in both directions without handing the sheet back from the catchers to the feeder's side [14, p. 683]. Lukens Iron Works became Lukens Iron and Steel Company (and later Lukens Steel).

The family was inquisitive and always looking for new technologies and ways to improve their product. For instance, "During a visit to St. Louis in 1883, Dr. Huston and his sons saw an early spinning machine in operation and were sufficiently impressed to order one for Lukens. It was put into operation in 1885 and could produce bowl-like shapes, called heads, up to seven feet in diameter and one inch thick [8]. Adoption of this innovation led to designing "manhole saddles," or covers for openings that allowed men to access the interior of a boiler to clean it. By creating shaped objects, Lukens Steel started to understand how the properties of steel changed with the methods of shaping. Later, they furthered these observations by scientific testing and reported the results to the ongoing international dialogue on testing and standards. When the technology of photography became available to them, Lukens used it to document

Figure 7. **Detail from Office Force (1895).** A. F. Huston, Dr. Charles Huston, and C. L. Huston occupy the center row. A. F. Huston ran the sales force and C. L. Huston was the works manager.

the mill, the employees, and later the microscopic structure of steel. Thus, Lukens was always a mixture of the old and the new: they used advanced communication technologies, such as photography, at the same time they used accounting books from the nineteenth century. In 1890 the plant was still a mixture of agriculture and industry as well; there was a truck garden for the workers next to the new mill (Figure 8).

In 1890 Lukens built the 120-inch three-high mill, at the time, the largest rolling mill in the country [15]. However, it soon became apparent that in order to remain competitive, they had to produce their own steel. Thus, after they opened the new 120-inch mill, they built two open-hearth furnaces for producing steel on the former site of the worker's garden. For rolling steel, the chemical composition was even more complex than for rolling iron, also the different-sized roll trains required different-shaped ingots; by making their own molds, they could also control of the size and shape of the ingot as well. Then, in 1899 they added a 48-inch universal plate mill, which produced plates with rolled edges from 9 inches to 48 inches wide and up to 100 feet long, and four more open-hearth furnaces [8]. All of this added complexity required more record keeping as they tried to track the amounts and chemical compositions of the ore, flux, and fuel;

Figure 8. **Lukens Iron Company (about 1880).** This photograph shows the workers' garden in the foreground and the new 120-inch mill, in the center rear, behind the 84-inch mill. Later this garden became the site of the first open hearth furnace at Lukens Steel.

track each batch of steel from pouring to inspection; fix machinery; and create new and modified machinery. The birth of companywide technical communication began at Lukens Steel at the end of the nineteenth century.

MODERNIZATION, 1900-1925

In 1898 Lukens Iron Works stopped rolling iron (after that they produced only steel) and in 1890 the firm incorporated. The original members of the firm were joined by long-time employees who took key positions in purchasing, production, accounting, sales, and operations. Dr. Huston died in 1897, but his sons entered the next century surrounded by loyal employees, sometimes several from the same family. Names such as Gordon (in sales), Humpton (as secretary, treasurer, and in testing), and Spackman (power and purchasing) were common, and some of these were on the board of directors as well. By the twentieth century multiple experts in many fields were necessary to produce Lukens' increasingly specialized product. The number of employees rose from 100 in the 1880s to over 2,000 by 1925 [16]. The majority of the foremen and managers were literate, many with both writing and drawing skills.

The physical growth of the plant during this time was immense. "In 1903, the company built a massive three-high, steam-driven mill with rolls 140 inches long that outclassed all its predecessors at that time, both at Lukens and in the nation" [8]. This mill was in a new building and had automatic roller-bed tables, another very important technical addition to modern steelmaking. It also incorporated Charles Lukens Huston's patented "Plate Straightening Machine," which became widely used by other steel-rolling mills as well [17]. One portion of the new mill building also housed five pit furnaces, each containing four pits to reheat the steel ingots preparatory to rolling [10]. By 1912 Lukens had also installed hydraulic presses in which heads could be formed over heavy dies, rather than by the spinning machine [8]. All of this increased complexity in technology was accompanied by an increased use of technical communication.

In 1917 Lukens began building what would become the largest rolling mill in the world [15]. In order to do so, they reincorporated and, at the same time, changed the name to "Lukens Steel Company." This 204-inch four-high mill was designed by Charles Lukens Huston and the United Engineering & Foundry of Pittsburgh. There was a great deal of correspondence between Huston and United Engineering, mainly in writing but also in drawing, as to how the mill should be built; this correspondence takes up several boxes in the archives. The big mill had a conveyor system that flipped the ingots for various passes through the rolls and eventually left them right-side-up for inspection. The main customer for this mill was the United States government who, during World War I, had an ever-increasing need for thicker and wider plate for marine boilers. However, when the mill rolled its first plate, the war was over and the shipping industry began to decline. The 204-inch mill remained the largest rolling mill in the world for many years and could roll plates up to 25 inches thick and 16 feet wide [13, p. 3]. Consequently, although the mill was a technological success, it caused financial problems. During the next several years the firm took out its first mortgage, closed parts of the plant, had layoffs, and stopped giving the shareholders dividends.

The end of the Lukens story, for the purposes of this book, came when they had to shift from the antiquated owner-operator business paradigm to a modern corporation. Company management became a difficult issue. Up to 2000 employees were operating in separate plants, under different divisions and foremen, and the reporting structures were weak. During this time Charles Lukens Huston used management consultants to try to get a grasp on the organization, but the split between Charles Lukens and Abram Francis Huston had created a lacunae in leadership wherein each individual foreman ran his own shop in his own way, and no one really knew how many employees there were or how much they were paid. The situation was not resolved until the board of directors voted that a comptroller be hired and the workers be paid from a central checking account. Abram Francis and Charles Lukens Huston retired in 1925 in the midst of this downturn, but Lukens Steel, built on a solid foundation, continued on as a family concern (Figure 9).

* * * * *

Lukens Steel was a family-owned corporation for over 188 years. For Rebecca Lukens, "Work and family were one" [2, p. 277]. That tradition continued through at least five generations. The Lukens plant survived the present downturn in the steel industry as well— the firm sold the plant to Bethlehem Steel in 1998, just before Bethlehem Steel went bankrupt in 2001, and then it was sold again. At the time of this publication, the plant is being operated by ArcelorMittal, the world's largest global steel corporation.

Since Lukens Steel was a family enterprise, the company kept many records that would have otherwise been discarded. Moreover, they valued them enough to deposit them at the Hagley Museum and Library for future generations to study. They left a valuable record of how communications changed between the nineteenth and twentieth centuries. The following chapters are organized according to when there were changes in the communication practices. They are separated into four time periods: 1810 to 1870, when the records were mainly accounting and correspondence; 1870 to 1900, when Lukens began to use record keeping to track their data; 1900-1915, when writing and drawing became a method of communication and problem solving; and 1915 to 1925, when technical communication increased so exponentially that it is difficult to analyze. Overall, the emergence of technical communication at Lukens Steel demonstrates how technical information became so complex that it had to be contained in

Figure 9. **Huston Homes and Plant.** The home of Isabella and Dr. Charles Huston is on the left, the home of Charles Lukens Huston and family is on the right and the plant is in the background. The photo was taken from the steps of Abram Francis Huston's home, Graystone. Courtesy Hagley Museum and Library, Acc. 50, Pictorial Collections.

documents rather than people. It was no longer possible to keep all of the knowledge necessary to the industrial process in a single mind; multiple minds had to work together. This is a major shift in human culture.

REFERENCES

1. J. C. Skaggs, *Lukens, 1850-1870: A Case Study in the Mid-Nineteenth Century American Iron Industry*, dissertation, University of Delaware, Wilmington, Delaware, 1975.
2. J. Scheffler, ". . . There Was Difficulty and Danger on Every Side": The Family and Business Leadership of Rebecca Lukens, *Pennsylvania History: A Journal of Mid-Atlantic Studies, 66,* pp. 276-310, Summer 1999.
3. *Rebecca Webb Lukens Bicentennial 1794-1994*, The Graystone Society, Coatesville, Pennsylvania, 1994.
4. United States Department of Transportation, Federal Highway Authority, *1795—The Philadelphia and Lancaster Turnpike Road,* accessed May 3, 2007 from http://www.fhwa.dot.gov/rakeman/1795.htm
5. G. G. Eggert, *The Iron Industry in Pennsylvania*, The Pennsylvania Historical Association, University Park, Pennsylvania, 1994.
6. A. C. Brown, *The Sheet Iron Steamboat Codorus: John Elgar and the First Metal Hull Vessel Built in the United States*, The Mariners' Museum, Newport News, Virginia, 1950.
7. R. B. Gordon, Personal communication to C. S. Johnson, June 6, 2007.
8. E. L. DiOrio, *Lukens: Remarkable Past—Promising Future,* Lukens Corporate Affairs Division, Coatesville, Pennsylvania, 1990.
9. G. J. Previts and B. D. Merino, *A History of Accountancy in the United States: The Cultural Significance of Accounting*, Ohio University Press, Columbus, Ohio, 1998.
10. J. S. Huston, "Vice President & Secretary J. Stewart Huston General Correspondence Lukens Steel Company History," B-2023, *Lukens Steel Archives*, Hagley Museum and Library, Wilmington, Delaware, 1944-1960.
11. R. A. Staples, *Reading Company*, accessed May 8, 2007 from http://www.thebluecomet.com/rdg.html
12. "Cost Accounting Charts for Steel Plant & Mills," B-183, *Lukens Steel Archives*, Hagley Museum and Library, Wilmington, Delaware, 1903.
13. C. T. Baer, *Lukens Steel Company Finding Guide*, Hagley Library and Museum, Wilmington, Delaware, 1994.
14. A. H. Fay, *A Glossary of the Mining and Mineral Industry*, United States Government Printing Office, Washington, D.C., 1947.
15. World's Largest Plate Mill, *The Iron Trade Review*, Chicago, pp. 1041-1045, November 23, 1916.
16. *A Century and a Quarter in Iron and Steel,* Lukens Steel Company, Coatesville, Pennsylvania, 1935.
17. C. L. Huston, "Plate Straightening Machine," United States Patent No. 712,300, 1902.

PART TWO

Analysis

CHAPTER 3

1810-1870: Prediscursive Technical Communication

Like many other industries in the early and mid-nineteenth centuries, Lukens Steel seldom used written technical communication: technical knowledge was transmitted directly from person to person, sometimes with body language, sometimes with the aid of spoken communication. In order for one worker to learn something from another, all he needed to do was watch; words would have been ancillary to this prediscursive form of communication, especially in a noisy factory environment. Prediscursive technical communication is that which takes place without written language—it can occur with spoken words or with body language alone. Also, in a business as small as the Brandywine Iron Works and Nail Factory, the workers were from similar linguistic backgrounds, so common cultural understandings prevailed and communication was easy; they were a close and stable group of individuals, working mainly in the same building and across generations. The plant itself was small: it consisted of a dam, raceway and waterwheel, the rolling mill, and a separate nail factory (Figure 1).

The need for written technical communication in the workplace did not yet exist. They kept a system of furnace journals, however, to record daily transactions, and Rebecca Lukens sold plate iron for boilers by writing letters. Most of the letters were bargaining for the price of materials, especially bar iron, but some contained technical specifications. During Rebecca Lukens' time, she saved incoming letters, but not outgoing ones. When Abraham Gibbons, Rebecca's son-in-law, joined the firm in 1842, he began keeping handwritten copies of outgoing letters as well. When Dr. Huston, Rebecca's second son-in-law joined the firm in 1849, they began keeping copies of all outgoing correspondence, including bills and receipts, in letterpress books. As more people joined the office staff in the 1880s, stamped forms and new handwriting appeared

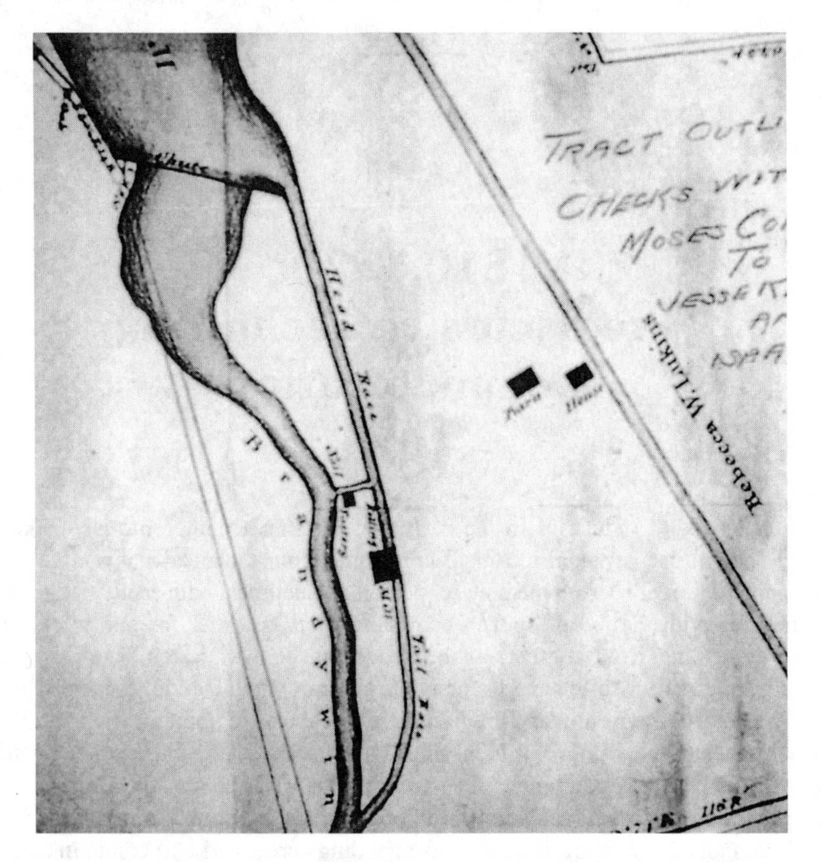

Figure 1. **Detail from a Map Rebecca Lukens' Property in 1810.**
The original Brandywine Iron Works and Nail Factory consisted of a dam, a raceway, and two factory buildings.

of the letterpress books, and outgoing communication became more complex. However, for the first 60 years at the Brandywine Iron Works, the forms of technical communication remained essentially static.

The following chapters about technical communication at Lukens Steel begin with a brief outline of the technological changes and processes in order to provide a context for the emerging forms of communication—changing technology increasingly required changing forms of technical communication. This chapter describes the system of financial record keeping in bound books—daybooks, journals, and ledgers—that lasted well into the twentieth century. Although these books do not contain technical communication, they were the foundation of record keeping. The only technical communication that existed in writing during this period was incoming and outgoing correspondence. The majority of this correspondence is preserved in letterpress books, on tissue-thin paper, in handwriting. During this period, 1810 to 1870, there were very few changes in the methods of technical communication at Lukens Steel.

THE TECHNOLOGICAL PROCESS

Between the years of 1810 and 1870, the machinery in a rolling mill was fairly simple. The Brandywine Iron Works and Nail Factory, built over the raceway for waterpower, had a waterwheel that powered the bellows for the reheating furnace, the anvil, shears, and rolls (Figure 2). Fuel was necessary to heat the iron before rolling and, until 1834 when the Philadelphia & Columbia Railroad was built, Lukens had to purchase coal from miles away carted by wagon. Similarly, pig iron billets were purchased from manufacturers and shipped by wagon. The finished material was then sent by cart to Philadelphia (38 miles) or Wilmington (26 miles) [1, p. 6]. The movement of raw material and product was slow, and the communication to and from customers was slow as well.

The technological process was, compared with later years, simple. First the wrought iron billets were heated so that they were soft enough to pass through the rolls. Lukens bought semifinished bar iron in billets (small) or blooms (large), which were then heated before processing. Sometimes scrap was added to the heating billets and the whole was pounded into a molten mass before rolling. "When heated thoroughly it was rolled out in the mill to pack it together and turned over in the rolls. It was then sent back to the furnace, right side up, heated thoroughly, and then rolled out into a plate" [2]. The workers used tongs to lift the iron, set it on casters, and fed it into the rolls. A "catcher" on the other side would grab the iron and pass it back to the front. A "screwman" would then adjust the distance between the rolls and the steps were repeated.

Although the process was simple, there were problems: sometimes they received brittle billets with poor chemical content. Rebecca Lukens continually negotiated, via mail, with her suppliers to get the best possible iron. Also, sometimes in a dry season there was not enough water to power the wheel. In this case, "the workmen would rush for the water wheel, climb up on its rim, and by their united weight help the 'pass' through the rolls, thus preventing a 'sticker,' which invariably meant fire-cracked rolls and, later on, broken ones" [1, p. 9]. At this point, the workforce was small—17 men headed by Solomon Lukens, Rebecca's brother-in-law—and the work was predictable.

The only changes in technology before 1870 were a new furnace, added in 1853, and in 1854 they enlarged the rolls to 60 inches. In order to eliminate the vibration from the waterwheel, they added a flywheel that could keep the power flow steady (Figure 3). This technology was all that was necessary for the Brandywine Iron Works and Nail Factory, and they continued successfully through the busy Civil War period simply by doubling the staff.

Figure 2. **Woodcut from Lukens Twentieth Century Stationery Depicting the Old Mill.** The anvil, shears, rolls, and furnace were powered by the raceway and waterwheel. At this time the mill employed 17 men.

Figure 3. **Flywheel for Power.** This flywheel was connected to the waterwheel to stabilize the vibrations and add more power in the 1850s.

DAYBOOKS, JOURNALS, AND LEDGERS

In general, accounting practices predated technical communication. In *The Visible Hand: The Managerial Revolution in American Business*, Alfred Chandler reminds us that, linguistically, the first word for a merchant's office was the "counting house" [3, p. 37]. In early American business there "was little uniformity in the accounting practices" and most accounts were kept in an "ad hoc, unsystematic way" [3, p. 70]. The Brandywine Iron Works and Nail Factory kept a systematic set of financial records, although they were seldom balanced. Dr. Lukens kept daybooks from 1816 (when he took over the mill), and the subsequent owners continued them until 1917 (the year they reincorporated). The company kept journals from 1813 to 1882 and ledgers from 1812 to 1938. In the later years these records were only a small subset of other financial records, but in the early years they, with the correspondence, were the only records that the company kept (Figure 4).

As described in Chapter 1, daybooks recorded daily transactions as they occurred, journals reorganized the daily data under individual or company names, separating them into credit and debit columns, and were updated approximately once per month, and ledgers summarized the year's transactions. These books

Figure 4. **Ledger, Journal, and Daybook (1820s).** This triad of account books was used by many ironworks in the nineteenth century—the large ledger (on the bottom) is the yearly summary, the journal a transitional book, and the daybook (on top) was for recording daily transactions. Lukens kept such books from 1810 to 1917.

were cross-referenced by giving each individual or company a number that was then written in the margins in all three books. This system of accounting was widely used but somewhat haphazard. In *Understanding and Using Early Nineteenth Century Account Books*, Christopher Densmore writes:

> Accounts were started in the ledger on the first available page. When there was no more room on a page to continue an account, the debit and credit sides were totaled and the sums posted to a new page. When there were no more available pages in a volume, a new volume was started. A single account might be carried through several ledgers—and many years—without being settled or balanced. Since ledger accounts were in no logical order, there was usually an alphabetical index with page references at the front or back of the volume [4, p. 8].

Some historical events are recorded in the daybooks, ledgers, and journals, such as the rolling of the first boiler plate in 1818 and the order for an iron-hulled ship, the *Codorous* in 1825. Also, Rebecca Lukens kept a journal called "Iron Book, R. W. Lukens & Co., Brandywine Iron Works" for the years 1842-1844. (Figure 5). These books were used frequently and had checkmarks in different colored pencil in the later volumes—someone went over them to check the entries and made notes as they did so. Rebecca Lukens' daughter Martha worked on the company books until she was unable to keep up with the firm's increased volume [5, p. 296]. Her husband, Abraham Gibbons, eventually withdrew from the firm to start a bank.

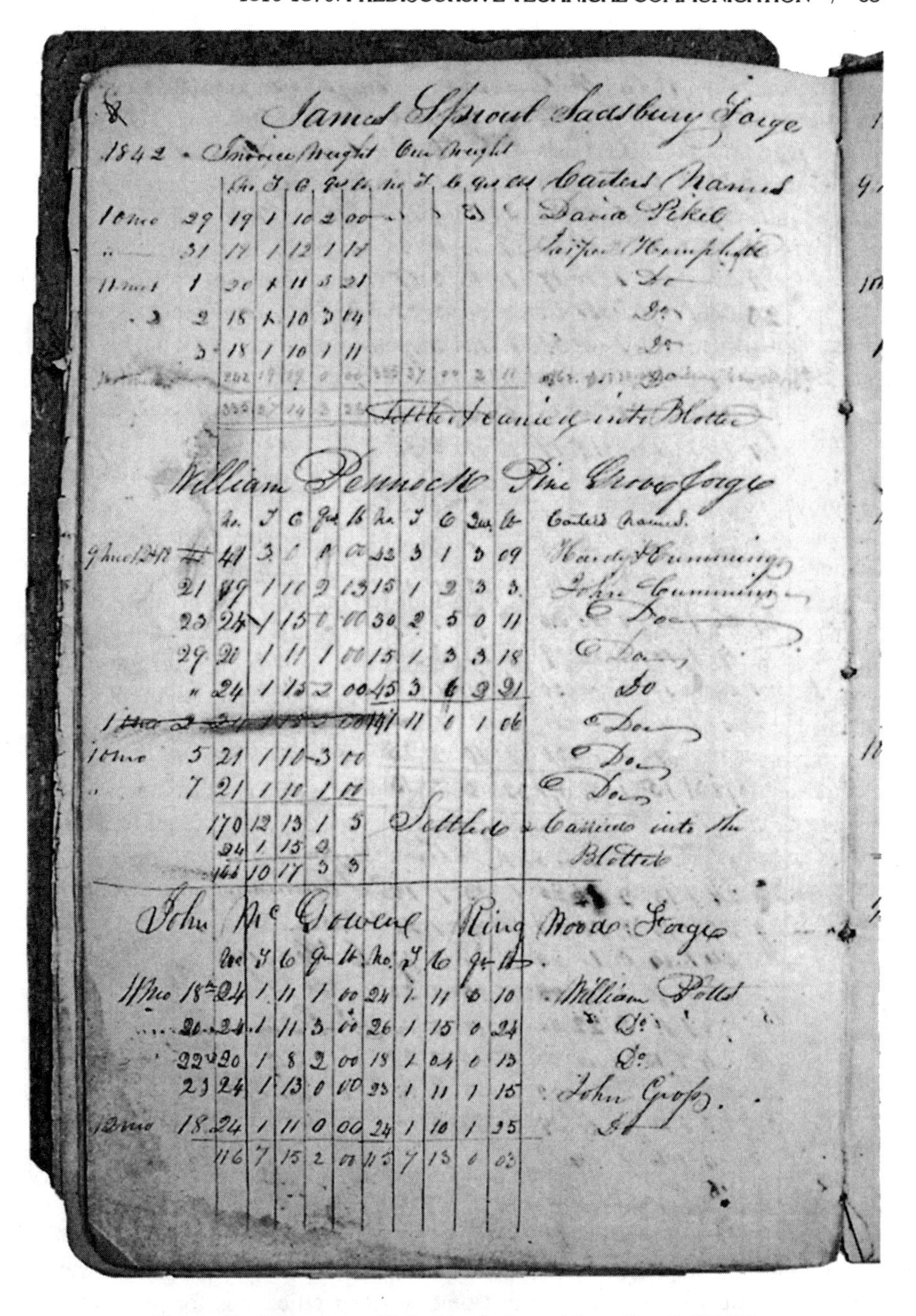

James Sproul Sadsbury Forge

1842 Invoice Weight Our Weight

Carters Names

David Pikel

Do

Do

Do

Settled & carried into Blotter

William Pennock Pine Grove Forge

Carters Names

Hardy & Cumming

John Cumming

Do

Do

Do

Do

Do

Do

Settled & Carried into the Blotter

John McGowen Ring Woods Forge

William Potts

Do

Do

John Gross

Do

Figure 5. **Page from Rebecca Lukens' *Iron Book* (1842).** Rebecca Lukens kept a separate journal for transactions between the years 1942 and 1944.

For an economist or accounting historian, the books have a rich history to tell. Julian Skaggs analyzed the books between the years 1850 and 1870 and was able to assess the firm's financial state throughout that time. This is true despite the fact that he found the books were never balanced or closed until December 1871. He surmised that this was probably because the partners were more interested in operating the business than examining its profitability [6, p. 177].

The firm used these account books for a very long time—from 1816 to 1938—even after the system had been joined by many other forms of record keeping. Other accounting books were cash books, trial balances, alphabetized transfer ledgers, voucher record books, sales journals, and of course, bank books. These are only some of the financial records, since the sales and shipping records were listed in a separate section. Thus, accounting practices predated technical communication. They are mentioned here because they provide a foundation for record keeping which, as the complexity of the technology and the number of employees increased, evolved into data collection and later, technical communication. In the case of the furnace journals, the books were later adapted to a variety of specialized uses. For instance, when the old mill (Brandywine Iron Works) was converted into a puddling mill for making steel in reverberatory furnaces, new journals were created, called "puddle mill books." Puddling was a job done by hand, so these are the last books that tie specific workers to their output. The puddle mill books included a list of the men repairing the furnaces, the names of workers and their positions, and daily and weekly tonnage separated into types. They had to break the page down into several sections to be able to track this information across time (Chap. 4, Figure 7). Other journals that emerged, which we will see in the next section, are car books, tonnage records, and defects books. They all bore a resemblance to the day books, journals, and ledgers, but increasingly the information they contained was organizational and technical, rather than purely accounting.

CORRESPONDENCE

Simple letter writing was a sufficient form of technical communication to run the business for 60 years. At first, incoming letters were kept folded in pigeon holes in cabinet desks (Figure 6). Although Rebecca Lukens saved her incoming correspondence, she did not make copies of her outgoing correspondence. Record keeping of transactions was based on mutual trust, and thus it was not necessary to retain letters as a record of dialogues. Many of Rebecca Lukens' suppliers, especially when she was just beginning to put the business on a solid foundation, were fellow Quakers who were not only honest but extended her credit as well.

The only technical writing at this time were specifications for plate iron. In the two pages of the following letter, one of her agents is ordering plates large enough for boiler heads (Figures 7 and 8). By convention, letters in this social discourse community began "Yours of the 22nd ins^{t}. came duly to hand." This

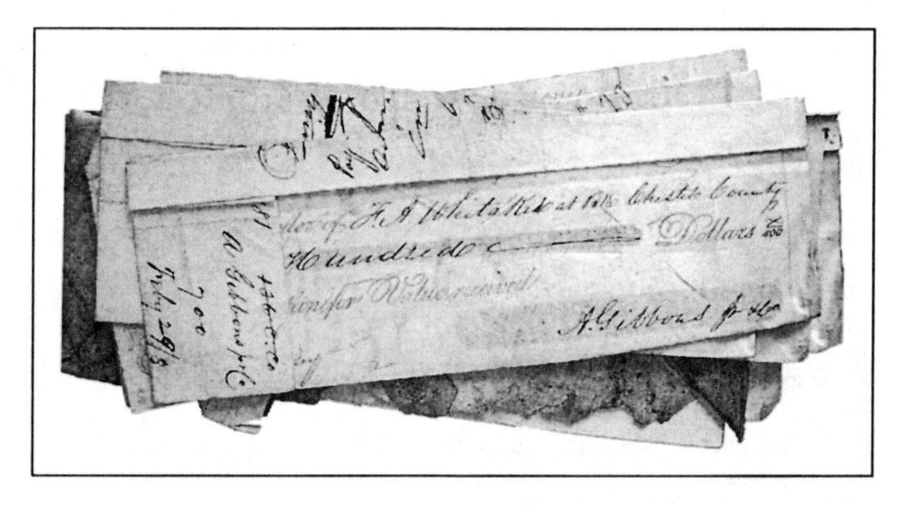

Figure 6. **Incoming Correspondence and Receipts.** In the early years of the firm, incoming letters and receipts were kept in pigeon holes in cabinet desks. Outgoing correspondence was not kept.

common practice lasted until late in the nineteenth century. Thus, the author of each letter began with a specific reference to the letter that it answers, ensuring an understanding of its place in the asynchronous (and slow) dialogue. The language in the letters was straightforward and clear, as it had to be because mistakes were expensive. Sometimes, at the beginning or end, if the writers were close associates they inquired after the family. However, overall the letters were short and linguistically to the point.

Abraham Gibbons began keeping handwritten copies of outgoing correspondence in a bound book in 1847. Dr. Huston joined the firm in 1849 and they began to keep copies of letters in letterpress books. The process for copying a letter into a press book is explained by JoAnne Yates in *Control Through Communication: The Rise of System in American Management*:

> A letter freshly written in a special copying ink was placed under a dampened page while the rest of the pages were protected by oilcloths. The book was then closed and the mechanical press screwed down tightly. The pressure and moisture caused an impression of the letter to be retained on the underside of the tissue sheet. This impression could then be read through the top of the thin paper [7, pp. 26-27] (Figure 9).

They used letterbooks not only to record their transactions and store their correspondence, but because "in legal circles it was accepted as a true copy of the original" [8].

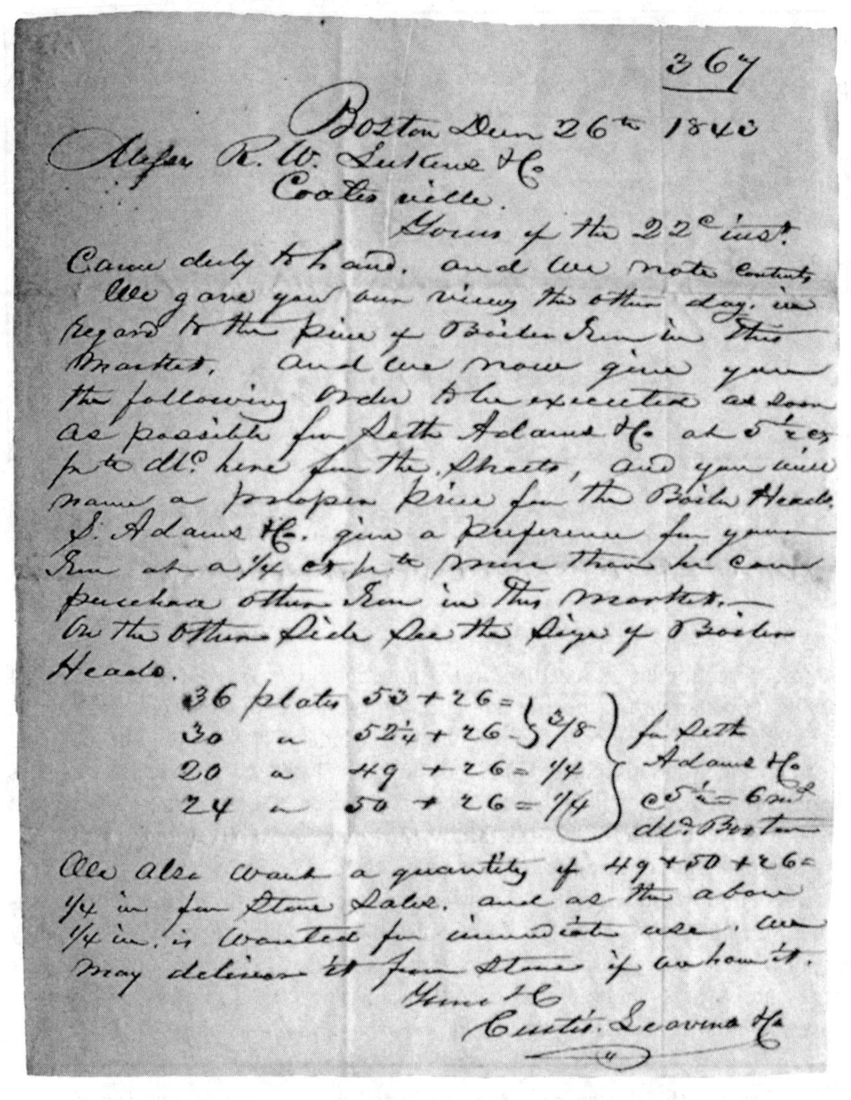

367

Boston Decr 26th 1843

Messrs R. W. Lukens & Co
Coatesville.

Yours of the 22d inst.
Came duly to hand, and we note contents
We gave you our views the other day, in regard to the price of Boiler Iron in this Market, and we now give you the following Order to be executed as soon as possible for Seth Adams & Co at $5\frac{1}{2}$ cts per lb here for the Sheets, and you will name a proper price for the Boiler Heads. S. Adams & Co. give a preference for your Iron at a 1/4 ct per lb more than he can purchase other Iron in this market.—
On the other side see the size of Boiler Heads.

36 plates	53 + 26 =	3/8	for Seth Adams & Co @ $5\frac{1}{2}$ – 6 cts dld Boston
30 "	$52\frac{1}{4}$ + 26 =	3/8	
20 "	49 + 26 =	1/4	
24 "	50 + 26 =	1/4	

We also want a quantity of 49 + 50 + 26 = 1/4 in for Store Sales. and as the above 1/4 in. is wanted for immediate use, we may deliver it from Store if we have it.

Yours &c
Curtis Leavens & Co

Figure 7. **First Page, Letter from Boston Agent (1843).** Lukens had agents in Philadelphia, New Orleans, New York, Albany, and Boston. These agents received technical specifications from customers and then wrote to the Brandywine Iron Works.

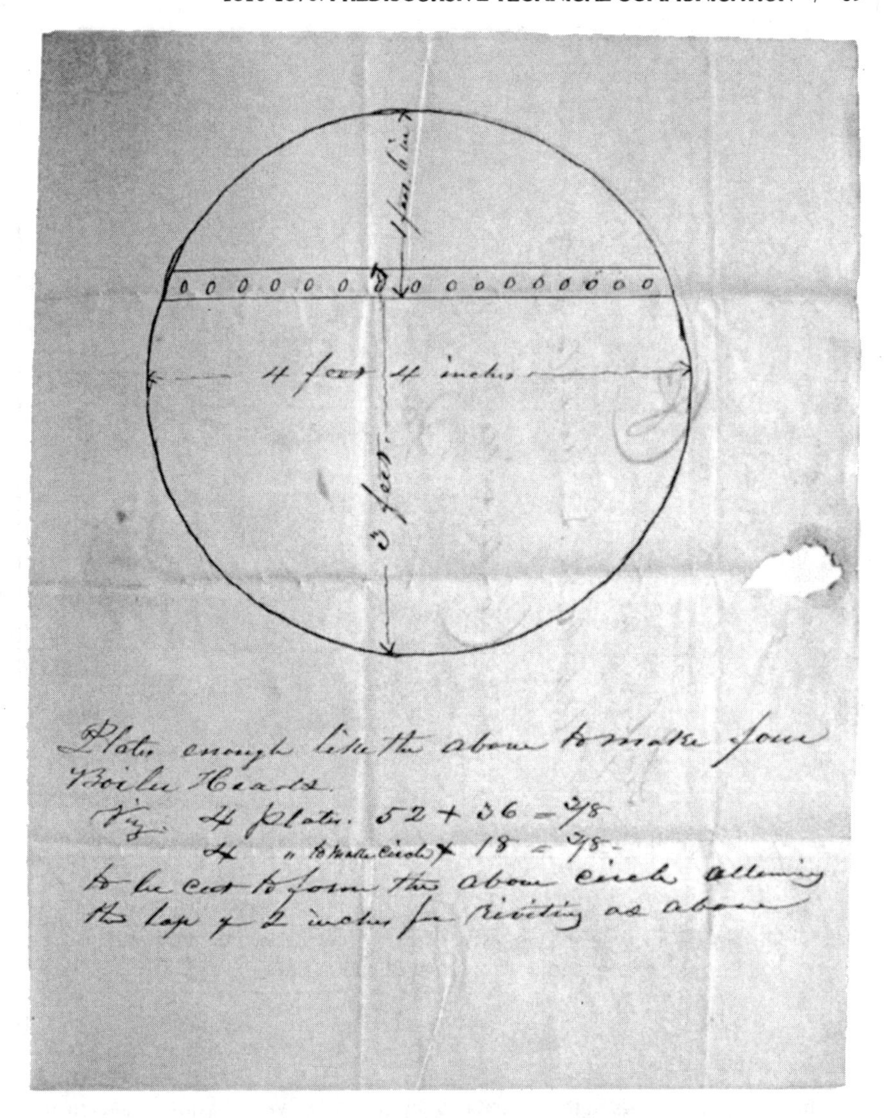

Plates enough like the above to make four
Boiler Heads.
Viz: 4 plates. 52 + 36 – 3/8
4 " to make circle + 18 – 3/8.
to be cut to form the above circle allowing
the lap of 2 inches for Riveting as above

Figure 8. **Second Page, Letter from Boston Agent (1843).** This drawing further described specifications for boiler head plates (to be placed at the end of a cylinder).

Figure 9. **Making a Letterpress Copy.** Each piece of correspondence, while the ink was still wet, was placed between two blotters and pressed onto a sheet of tissue paper.

The letterpress books in the Lukens collection are of varying quality, according to letterbook manufacturer, ink used, writer, and the person pressing the book. Some are clear and legible, others are faded and nearly impossible to read. Some have watermarks at the edge of the page and in some the ink ran, making a blurry image. Since the ink from the letters was transferred to the back of the tissue paper, sometimes it is more legible from the back, although the text was backwards (Figure 10). Thus, letterpress book pages are difficult to read without inserting white paper between pages.

Frequently the firm used "Mann's Parchment Copying Paper" books, which gave extended directions on the inside front cover for making copies of many

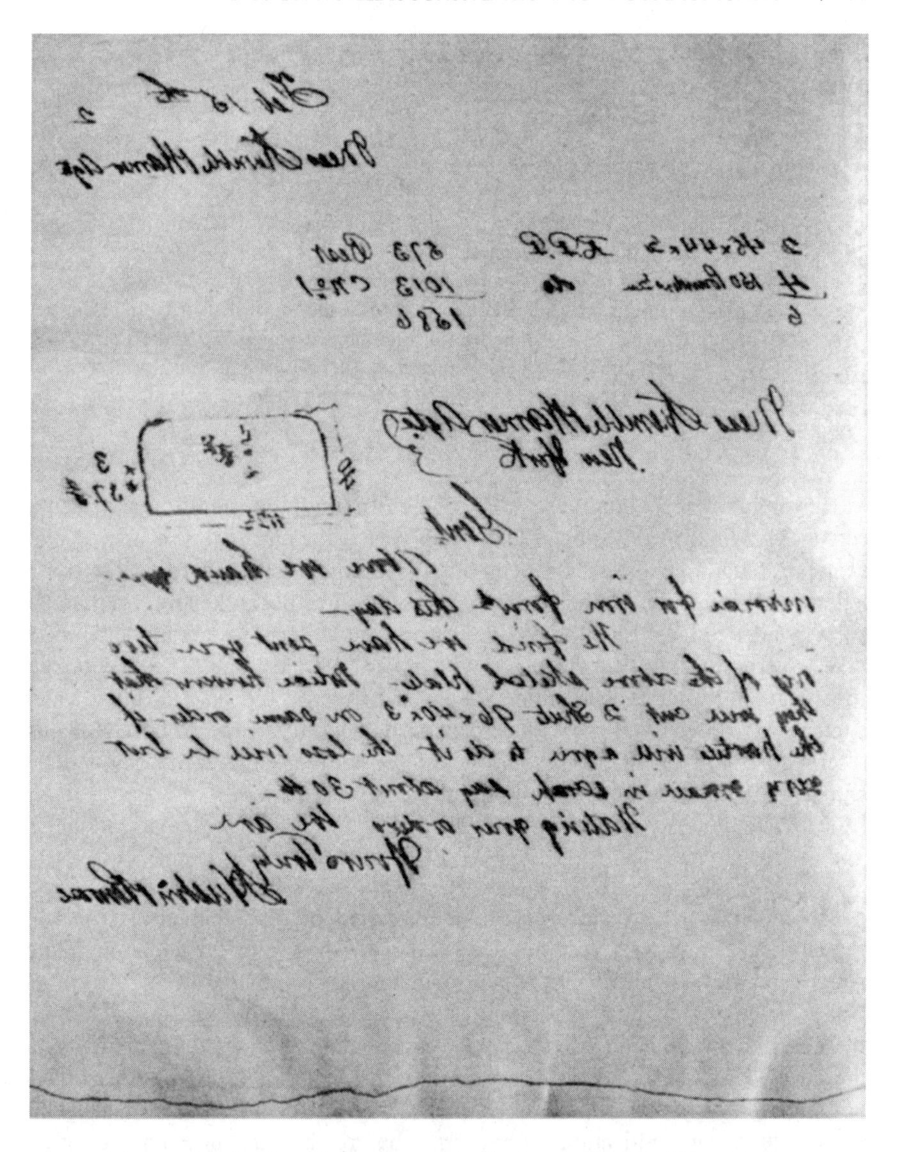

Figure 10. **Back of Page from a Letterbook (1862).** Although the back of the page is legible, the image on the front of the tissue paper is faint.

letters at the same time, as must have frequently happened as the volume of correspondence increased.

> Procure a tin box with lid to hold 20 blotting pads.
>
> Dip half the lot of blotters in water, let them drain off a few moments; then place a dry blotter between each wet one, give them a few minutes squeeze in Press and they will remain wet for three days; then take an oil sheet, place it to the left; then lay a wet blotter; then turn leaf of Copying over on blotter; then lay your letter on; then another oil sheet, and so on, and you can copy all your letters at one time, thereby saving time. With a little care and experience as to wetting at first, you will be so well pleased, as never to go back to old way [9].

This ungainly technology was used by the firm until the end of the nineteenth century. At first, one 700-page letterbook was sufficient to record one or two years of outgoing correspondence, but that number climbed to 10 or more letterpress books per year until 1903, when the firm began keeping individual correspondence files. Like the earlier handwritten letters, those in the letterpress books were short and to the point with no excess language. They contained specifications, receipts, bargaining for lower prices on blooms and billets, and attempts to resolve problems with defective iron. During her life, Rebecca Lukens reserved her emotional and creative writing for her journals and personal letters, and her daughters and sons-in-laws did the same.

* * * * *

This chapter, which covers the longest period of four chapters (60 years) is significant in that almost nothing changed. This is true even though the American railroad network was reaching across the country, bringing in its wake the telegraph and faster mail service. "In 1849 freight moving from Philadelphia to Chicago had to pass through at least nine transshipments in the course of as many weeks: ten years later the journey took only three days and required only one shipment" [3, p. 112]. As Yates wrote, there was "very little written communication within firms before the late nineteenth century, and oral communication was predominantly informal and undocumented" [7, p. 65]. Prediscursive technical communication—oral and physical—was all that was necessary to this comparatively simple manufacturing process. The technology was stable, and the methods of communication were simple during the 60 years covered in this chapter. The following chapters of this book, covering 30 years, 15 years, and 10 years respectively, show so much change that it is difficult to keep up with the emerging forms of technical communication.

REFERENCES

1. *Handbook of Products*, Lukens Iron and Steel Company, Coatesville, Pennsylvania, 1912.
2. J. S. Huston, "Vice President & Secretary J. Stewart Huston General Correspondence Lukens Steel Company History," B-2023, *Lukens Steel Archives*, Hagley Museum and Library, Wilmington, Delaware, 1944-1960.
3. A. D. Chandler, *The Visible Hand: The Managerial Revolution in American Business*, Harvard University Press, Cambridge, Massachusetts, 1977.
4. C. Densmore, Understanding and Using Early Nineteenth Century Account Books, *The Midwestern Archivist, 5*:1, pp. 5-19, 1980.
5. J. Scheffler, . . . There Was Difficulty and Danger on Every Side": The Family and Business Leadership of Rebecca Lukens, *Pennsylvania History: A Journal of Mid-Atlantic Studies, 66,* pp. 276-310, Summer 1999.
6. J. C. Skaggs, *Lukens, 1850-1870: A Case Study in the Mid-Nineteenth Century American Iron Industry,* dissertation in history, University of Delaware, Wilmington, Delaware, 1975.
7. J. Yates, *Control through Communication: The Rise of System in American Management*, Johns Hopkins University Press, Baltimore, Maryland, 1989.
8. M. Hodgson, *Conservation Distlist—Letterbooks*, 2004, accessed September 8, 2007 from http://palimpsest.stanford.edu/byform/mailing-lists/cdl/2004/1259.html
9. Letterbook, B-1214, *Lukens Steel Archives*, Hagley Museum and Library, Wilmington, Delaware, 1879.

CHAPTER 4

1870-1900: Record Keeping Paves the Way

After Lukens built the first steam-powered mill in 1870, the company moved from the prediscursive world of technical communication into a culture of record keeping that was made necessary by the complexity of the new machinery. First, they began to use car record books, puddle-mill books, bound inventories, and payroll sheets. After the 120-inch mill and the two open-hearth furnaces were added in 1890, they also began to keep records of open hearth and plate mill output, records of defective plates, and some testing and inspection documentation. Some of these record-keeping systems began with individual foremen who kept notes in pocket books or on single sheets of paper and then brought them to the office where their data was added to ledgers. This record keeping paved the way for further use of writing; it was a quantitative literacy and a use of paper on the factory floor that became more and more common as the technology became more and more complex. When the foremen and workers in various parts of the factories had to keep lists of the products they made and the defects they saw; when the men in charge of the railroad cars coming into and leaving the plant had to carry notebooks listing dates, incoming supplies, and outgoing product; when even the puddle-mill foremen kept personal notebooks of tonnage, writing began to be a tool that was integrated into the factory environment. It also became a dataset for further analysis as the data was handed up the managerial scale, compiled into bound books, and analyzed for patterns. Quantitative literacy was the first step toward more widespread literacy in reading, writing, and drawing.

Before 1900 there was still little of what we consider technical communication today, such as test reports, procedures, manufacturing plans, and manuals. There was little evidence of testing, even though Dr. Huston published two papers reporting the outcomes of tests on the properties of iron in the *Journal of the Franklin Institute*. This chapter shows how record keeping was the first written and saved documentation at Lukens Steel. The accounting and correspondence described in the previous chapter continued on as before, in addition to this new record keeping. There was only one advance in this realm: Lukens bought their first typewriter. From that point forward some of the letters in the letterpress books were typed rather than written. This technology helped to hasten knowledge transfer between Lukens and the companies to whom the letters were sent, since handwriting, especially that of Charles Lukens Huston, works manager, was hard to read. The typewriter stayed in the main office and was not used for intraplant communications until after 1910.

This chapter will first give an overview of the new technology at Lukens Steel that led to the development of new record keeping systems. After an overview of the technological changes, each new genre will be described: car record books, puddle-mill records, tonnage records, defective records, inventories, and payroll sheets. The chapter will then discuss Dr. Huston's participation in the new world of testing and the development of standards. Also, Lukens Steel had been collecting patents and, although they were collected and kept mainly for research purposes, they will be discussed here. The greatest advance for technical communication at this time other than record keeping, came with the introduction of the typewriter. This set the stage for an increased rate of knowledge transfer by making accurate and detailed information exchange possible.

THE TECHNOLOGICAL PROCESS, 1870-1900

After the Lukens Iron Works reaped great profits in the Civil War, they built an 84-inch steam-powered mill, finally taking advantage of boiler technology, and the old mill was converted to a puddling mill (Figure 1). Puddling required skilled, heavy labor working iron in reverberatory furnaces. In a puddling furnace, mineral fuel was kept out of contact with the metal to avoid contaminating it with sulphur—the fuel was on one side of the furnace, separated by a brick wall and the iron on the other. In the process the carbon in the pig iron was removed by oxidation to make carbon-free wrought iron. The workers accessed the hearth of the furnace by a door, loaded pigs and scrap, waited until some turned red, shifted them around until they all turned white and melted, and then broke and stirred them into balls 12 to 15 inches in diameter. During this process many impurities burned out. When the iron solidified, the puddlers could feel it, which was called "coming to nature" [1, p. 49]. Producing their own billets lessened Lukens' reliance on outside vendors, which gave them additional control over their product. Puddlers were expert craftsmen who were paid by the ton, so records of the puddlers, their assistants, and their output had to be kept. The accounting books mentioned in the last chapter were easily adapted into puddle-mill journals in which all this data was kept.

The Wilmington and Northern Railway built a railroad spur directly to Lukens Iron Works in 1869, so receiving supplies and shipping product became easier than ever. The new 84-inch mill could roll wider and longer plates for the ever-increasing demand in the shipping industry (Figure 2). This new mill

Figure 1. **The Old Mill Converted to a Puddling Mill.** Abram Frank Huston, Charles Lukens Huston, Dr. Charles Huston, Joseph Humpton, and H. B. Spackman are in front of the waterwheel.

immediately began selling less expensive, wide boat iron [2, p. 173]. By then, stamping the tensile strength on sheet iron for railroad and marine engine boiler plate became mandatory. Iron leaving the plant had to pass inspection and therefore, inspection came to be part of the manufacturing process. In 1875 Dr. Huston purchased a testing machine and became active in forming industry standards. However, even though he performed experiments, published the results, and worked with others to formulate some of the first boiler-plate industry standards, he did not keep any records of this work, nor were any kept permanently in the office. The company was still in a transition from prediscursive to discursive technical communication.

In 1890 Lukens underwent another major expansion. They built a 120-inch, three-high, roughing and finishing mill for rolling steel plate. It had power operating tables that were tilted up and down by a hydraulic mechanism and were driven by 1300 hp Corliss steam engines. A set of straightening rolls was added in 1898 for leveling the plates as they came out of the rolling mill while they were still hot, a process that Charles Lukens Huston patented; subsequently its

Figure 2. **84-inch Steam Powered Mill.** This mill, built in 1870, was the first significant expansion that Lukens made. It was later dwarfed by the 204-inch mill opened in 1918 (Chap. 6, Figure 2).

use was widely adopted [3]. This new mill was a major accomplishment and Lukens had it photographed the day it opened (Figure 3). On the site of the former worker's garden, two open hearth furnaces were built. Lukens could now have control over the quality of the steel ingots and therefore, control over the end product (Figure 4). However, as they found, it was difficult to make quality steel ingots, and they had to institute a system of record keeping to help them to do so.

The rolling mill itself was a complex series of processes, many of which could fail. As Alexander Holly wrote:

> The rolling-mill is, throughout, a series of machines, and much of their work is of the most difficult character . . . . It involved the adaptation of steam engines and boilers under peculiar environment; of roll trains, which are by themselves a vast department of engineering; of power-handling, finishing and transportation, and of the utilization of fuel under varying circumstances and on a gigantic scale [4].

Open hearth furnaces were equally complex. The chemical content of the pig iron (and other chemical ingredients), the physical condition of the fuel, the

Figure 3. **Start-up of Lukens 120-inch Mill (1890).** This rolling mill, along with two open hearth furnaces, marked the beginning of complex technology at Lukens Steel.

Figure 4. **Open Hearth Furnace #1.** Open hearth technology was dependent on invisible chemical reactions that changed according to the contents of the pig iron, scrap, fuel, and chemicals they used, along with the temperature of the fire, the length of the heat, and the size, shape, and condition of the mold it was poured into.

brickwork supporting the furnace and holding the hot air, the layout of the "soaking pits" where the ingots cooled, and shape of the molds all affected the quality of the ingots. Getting ingots that were the right chemical composition, size, and shape, without defects, was an ongoing battle with unseen chemical and physical processes. The size and quality of the ingot affected the outcome while it underwent rolling and any further work, such as flanging, punching, or drilling. The quality of the ingot also affected its strength and durability. Since the final product was often subject to government or railroad inspection, there was no room for error. Each batch of steel was an experiment and Lukens could never be sure of a high-quality finished product. The difference between them and their competitors was that Lukens never stopped trying to analyze the processes. The first step in this ongoing analysis was to keep written records.

During this time, the Lukens Steel plant went from a semi-pastoral to an industrial setting: for instance, the worker's truck garden, as seen in Chapter 2, became the site for Open Hearth #1 (Chap. 2, Figure 8). In 1899 four more open hearth furnaces were built on the other side of the Brandywine and the plant spread out from its original core. In that same year a 48-inch universal, three-high rolling mill was also built to make thin, wide steel next to the new open hearth furnaces. All of these new furnaces and mills had buildings for storage, gas-, electric-, or coal-powered generators; offices; lunch rooms; toilets; locker rooms; separate storage units for items such as brick, patterns, sand, hose, ice, rivets, locomotive repair, and small structures for the various watchmen. In addition, Lukens was still providing some worker housing. What had been a comparatively simple plant became a distributed set of buildings, each with its own complex technology and outbuildings, spread over a large area.

NEW FORMS OF DATA COLLECTION AND RECORD KEEPING

The records discussed in this section are data collection, not technical communication. However, they led to an increased reliance on paper to document and store data and a need for an increasingly large staff of clerks to manage it. They also demonstrate that quantitative literacy had evolved, from necessity, on the factory floor, since many of the records originated from foremen, kept in small books or on sheets of paper. The car record books were essential for keeping track of incoming materials, outgoing product, and costs incurred on different railroads. The puddle-mill books were a form of journal, which recorded daily information about the workers, tonnage, repairs, and other activities. The tonnage and defective records were essential in collecting data to control the quality and quantity of steel in both the open hearth furnaces where it was made and the plate mills where it was rolled. The other miscellaneous record keeping, such as inventories and payroll sheets, show an increasing, if haphazard, attempt to rise above the individual memory and create a corporate one [5, 6].

Car Record Books

Car record books were kept from 1874 to 1897. From them we can see what materials were arriving by train. Lukens kept a careful accounting, not only for financial and inventory reasons, but because demurrage (cars blocking the tracks with unloaded materials) could be costly. The small books carried by the men document the railroad from which it originated, the car number, what time it arrived, what it held, what time it left, and the contents when it left (Figure 5). They moved vast amounts of materials by railroad car, especially prior to electric cranes. They shipped forgings, castings, coal, brass, brick, molds, pig iron, bar iron, scrap, rail ends, steel, a milling machine, clay, pipe, springs, and more. The car record books were wide rectangular books, bound in black, with the words "Car Record" and "L. I. & S. Co." imprinted in gold on the cover. The entries are neat and complete and were copied from the rougher notes, kept in small books, and taken by the labor boss in charge of loading and unloading (Figure 6).

The car record books are significant in that, for the first time, some of the workers and foremen were given writing tools and the responsibility to take notes throughout the day. They demonstrate that paper was finally being used by the workmen to control the plant itself. That the workers welcomed this new responsibility can be seen in the ornate cover in Figure 5 and in the other handwriting in the small volumes.

Puddle-Mill Records

The puddle-mill books were a natural outgrowth of the furnace journal, keeping new types of information in a slightly different format. Puddling was still a workman-based industry, dependent on the knowledge of skilled individuals, and thus they were paid in accordance with the tonnage they produced. The books, kept by a single person, listed each furnace, the team that worked it, and its daily output (Figure 7). The puddling teams were able to accomplish about five heats per day. Each page of the book covered a period of five days. The left side of the page often included remarks regarding miscellaneous jobs performed by puddlers (such as repairing a furnace or working in the ice house) and the number of helpers paid. Soon they began keeping rolling mill information in these books as well—there is a separate section in the lower left-hand corner naming the heater, helper, screw, roller, catcher, leverman, stocker, and then the type and number of blooms. The puddling journals changed across time and contained information that did not fit into the financial records, correspondence, or payroll sheets.

The puddle-mill records demonstrate how one genre, the furnace journal, was modified to meet the needs of a specific technology. It was necessary to keep track of how much iron was made, and no genre existed to fill this need. Thus, they adapted an old genre to fill a new need. The puddle-mill records are also significant in that they are the last tonnage records that reflected individual

Figure 5. **Chas. S. Thompson's Car Record Book (1897).** This small (approximately three by five inches) car record book contains much of the same information as in the car record book in Figure 6, but in a much rougher state, with dirt, corrections, and wear.

1888

Initial	No. of Car	When Received	Contents	When Sent Out	Contents
W. N	456	3 mo. 29		3 mo. 29	[illegible]
P. R	19233	" "		" "	Cinder
"	5209	" "		" "	"
"	5058	" "		" "	"
L. V.	353	" "	Gen. 18 tons [illegible] Pig	31	Empty
P. R	1007	" "	Robt. Moffly Co. 24920 cast scrap	31	Empty
L. V.	3964	" "	Kissinger & Yocum 22900 "	4 mo. 2	"
W. N	7	30		3 mo. 30	J. F. O. D.
P. R	3812	" "		" "	D. H.
N. C.	11404	" "		31	M. 1187½ - Fire. [illegible]
P. R. R	36835	" "	J. H. Weaver & Co. 21.17 [illegible]	4 mo. 2	P. M. C. L.
P. R	3708	" "		3 mo. 30	M. 1207 - 985½
"	2019	" "		" "	M. 1289
"	3520	" "		" "	W. B. [illegible]
"	3206	" "	Naylor & Co. [illegible] Steel	31	D. N. L. E. & Co.
C. M. & B	3036	" "	do. [illegible] "	" "	M. 1240 [illegible] F. 1215
P. R	40821	" "		" "	Cinder
"	23570	" "		" "	"
"	15293	" "		" "	"
E. & B.	135	31	Chs. Brooke & Co. 22 tons Pig	4 mo. 2	Empty
"	115	" "	do. } 30 tons "	" "	"
"	127	" "	do.	" "	"
"	132	" "	do. 13 tons "	" "	"
B. O	24091 x	" "	J. H. Weaver & Co. 21.12 Coal x	3	Empty
C. St. L. P	5777	" "	Naylor & Co. move of Steel	" "	"
Emp. Line	141	" "	do. [illegible] "	" "	"
B. O	24997 x	" "	J. H. Weaver & Co. 21.06 Coal x	4	"
P. R	4520	4 mo. 2	C. Ridgway & Son - Rebated Pig	3	Empty
P. R. R	17174	" "	Poulterer & Co. 24170 Cast scrap	" "	"
"	9248	3		" "	F. 8199 - 8150
"	17699	" "		5	M. 1367½ - 1247
C. R. R	38139	" "	Condemned plts.	3	W. B. [illegible]
P. R	0180	" "	Penna. S. Co. 28405 Steel	4	[illegible]
E. T. V. & G.	4137	" "	Naylor & Co. move of "	" "	Empty
P. R. R	40492	" "	do. [illegible] "	" "	"
N. Y. & N. E.	121849	" "	Hatfield & Phillips [illegible] Fire Clay	" "	"
B. O	23146 x	4	J. H. Weaver & Co. 21.15 Coal x	5	"
"	24697 x	" "	do. 21.06 " x	7	"
N. Y. & N. E.	12056	" "	Naylor & Co. move of Steel	5	"
[illegible]	3392	" "	do. [illegible] "	9	"
P. R	3323	5	Penna. Steel Co. [illegible] "	6	J. N. C. Co. M. 1325
L. V.	7698	" "	Kissinger & Yocum [illegible] cast scrap	10	Empty
P. R	4202	" "		6	F. 1215 - H. C. 1283

Figure 6. **Page from Car Record Book (1888).** From 1874 to 1897, employees at Lukens kept record books of shipments arriving and leaving by rail. This nearly copied version, in a book approximately 14 by 18 inches, was made from individual notes such as the book in Figure 5.

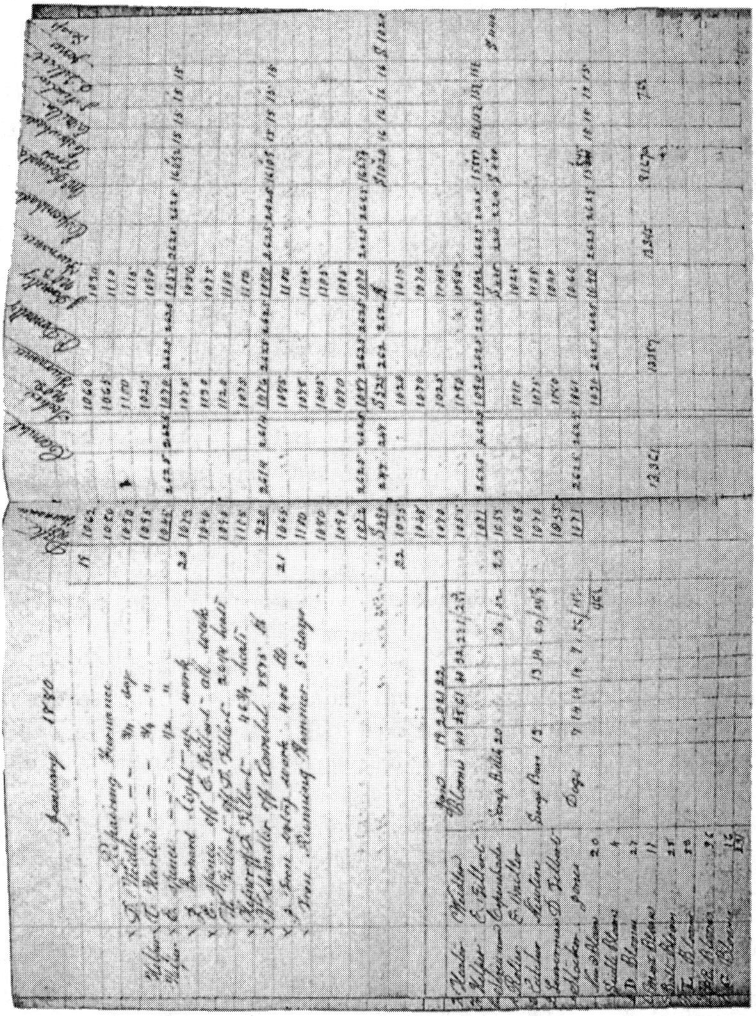

Figure 7. **Two Pages from Puddle-Mill Journals (1880).** The large-format pages listed each furnace, the team that worked it, and its daily output. They are the last production records that name individual workers.

output; after the open hearth furnaces were fully operating, tonnage of steel made was listed under the furnace, rather than the men who worked it.

Tonnage Records

In the puddle-mill records, men and technology were still one. However, after the opening of the open hearth steel furnaces in 1891, Lukens began keeping a series of books for both open hearth tonnage and plate mill tonnage, which were organized not by people performing work, such as in the puddle-mill records, but by physical and chemical aspects of the product and the output. The line of development was not straight: some of the tonnage books still kept narrative information about events and tonnage, such as the pocket book carried by a foreman, who listed events at the plant including deaths, downtime, fires, and other significant events (Figure 8). Most of the tonnage records, however, were sheets of raw numbers (Figure 9). All of these records were collected, brought into the main office, and entered into hardcover ledgers. They were important not only to keep track of the factory output but because the men were still paid, in part, by the ton.

Three forms of the hardbound tonnage documents exist in the Lukens archives. The plate-mill "Tonnage Ledgers" were kept from 1891 to 1918 and at first, were preprinted ledgers with columns for ingots, slabs, and flue plates, further broken down into weight charged, weight trimmed, and percentage yield. They are organized by mill, with separate sections for the 120-inch mill and the 84-inch mill. The master "Open Hearth Heat" books, kept from 1891 to 1901, were bound in black and recorded each heat number, the type of pig and scrap, the weight of each, and then a chemical analysis of the results. Often the amounts were corrected by red pencil. These records were significant because the heat number followed the ingot through the factory process and through rolling so that if there were defects, the plate could be tracked back to its source. Also, some of the workers were still paid by tonnage, so the tonnage records from the factories, the weigh stations, and the office had to match.

The whole process of making faultless rolled steel for boiler plate, much of which had to pass inspection, was extremely complex, and these records helped when it came to the tracking of defects. Later, the testing department added a form called the "Report of Tests of Steel," which also aided in the tracking of defects. These tonnage records continued until 1918, but by then they had evolved into indexed journals of the tonnage going to specific customers, a different use for a preexisting format. Written records evolve as the need for them evolves.

Defective Records

Some of the most important floor documents that Lukens kept were the open hearth defective records. Many of the foremen kept extensive notes of various problems, some of which originated in the open hearth furnaces, some of which

120" mill standing all week – increasing Slack up
" " off 2 days – Taking stock. →

84" mill off 2 days – Taking Stock. →

[illegible] Housing on Big Mill 12/7/95 – about 6.30 p.m.
Started up 12/22/95 A.M.

120" mill standing from 6 a.m. 2/7/96 to 6 p.m. 2/10/96
84" mill from 6 a.m. 2/7/96 to 6 a.m. 2/11/96
} on acct. of death of Mr. Calley →

Week Ending	Iron	Steel	Total
Amt. Fwd.	24 210	58 579 895	58 604 105
Sept 7, '95		1 393 034	1 393 034
" 14, "		1 228 878	1 228 878
" 21, "	4 469	1 293 153	1 297 622
" 28, "		1 452 055	1 452 055
Oct. 5, "		1 534 580	1 534 580
" 12, "	4 958	1 626 740	1 631 698
" 19, "	1 229	1 350 493	1 351 722
" 26, "		703 772	703 772
[illegible]	[illegible] 2 188	[illegible]	[illegible]
	37 054	70 033 221	70 070 275
1895			
Nov. 2, 1895		350 627	350 627
" 9 "		1 432 582	1 432 582
" 16 "	444	1 824 840	1 825 284
" 23 "	837	1 737 661	1 738 498
" 30 "		703 558	703 558
Dec. 7 "		809 233	809 233
" 14 "		1 129 295	1 129 295
" 21 "	4 244	778 391	782 635
" 28 "		1 142 578	1 142 578
1896 Jany 4, 1896		1 202 558	1 202 558
" 11 "		1 002 298	1 002 298
" 18 "	953	1 272 208	1 273 161
" 25 "		1 637 632	1 637 632
Feb. 1 "	17 515	1 439 020	1 456 535
" 8 "		1 061 971	1 061 971
" 15 "	2 101	1 140 412	1 142 513
	26 094	18 664 864	18 690 958

Figure 8. **Individual Plate Mill Weekly Output Book (1893 to 1900).** This journal, small enough to fit in a pocket, was a step in the evolution from the early nineteenth century journals to the later complex mill output record-keeping system. It recorded tonnage details as well as noting significant events, such as the day the 84-inch mill burned and the deaths of various workers, including Dr. Charles Huston in 1897.

Tonnage 6 pm		Weekly Output		Daily	
Date 1899	Tonnage	84" mill	120" mill	Entered	Output
Nov 1	4226759		[illegible]	617209	575632
2	4320578	685412	[illegible]	469506	565396
3	4071060	3422.7	1001.6	294532	544600
4	4224039			[illegible]	[illegible]
Nov 6	[illegible]			[illegible]	[illegible]
7	4076504		3065671	[illegible]	[illegible]
8	3903196	709141	2356530	139101	621707
9	[illegible]	3525	[illegible]	298673	[illegible]
10	3328166			310783	411903
11	[illegible]			251611	500406
Nov 13	3566812		[illegible]	102308	285640
14	4365105			106328	564457
15	4298700	644029	1992424	446365	512776
16	4568405	3220.0	946.2	518930	559225
17	4176941				479829
18				832362	299993
Nov 20	4487528		1947771	429389	198802
21	4578761			585622	417354
22	4657274	735519	1212276	355650	277137
23	4705995	3677	606.1	436558	357859
24	5314647			712106	[illegible]
25	5252804			432893	409176
Nov 27	5303358		[illegible]	597111	297757
28	5566492			538429	527495
29	5131573	690095	2267260	137843	572742
30		3450	1133.6		429127
Dec 1	4703130			455749	485085
2	4118859			126653	445249
Dec 4	4106940		2700908	101832	177916
5	3757949			246787	561781
6	3498740	630137	2070771	189260	442469
7	3146480	3150	1035.5	161274	513634
8	3912587			362856	496799
9	2907890			501762	506409
Dec 11	3313507		2922532	671182	265565
12	3195328			298431	416610
13	3067723	738864	2183665	374446	612446
14	2928455	3694	1091.8	407206	546651
15	2993116			529072	464409
16	2611868			336663	716021
Dec 18	2686603		681261	166014	89269
19	3850950			380307	167960
20	2747747	566476	114785	262230	135423
21	3244695	2832	57.3	612374	139426
22	3402304			236792	149180
23	3449939			197659	
Dec 25	off		476727		
26	3098763			499426	
27	4867623	326806	1499723	188860	
28	4426601	1634	749	140717	81734
29	5078355			285009	106219
30	5175367			395255	285774
Jan 1			2663989		142810
2	5838739			323952	977695
3	5518302	645564	2018416	196757	517224
4	5085354	3227	1009.2	142908	572801
5	5057426			438927	469885
6	5508009			981153	633570
Jan 8	5444691		3178564	244041	312359
9	4988876			93569	540984
10	5052563	652659	2426905	616008	652412
11	4560280	3263	1210.4	72803	565686
12	4188696			188931	560515
13	4713706			564318	539807
Jan 15	4161469			154700	205957
16	3162302			186351	522995
17	3419196			703655	436664

Figure 9. **Loose Tonnage Sheets (1899).** These loose sheets represent an early stage in the record-keeping process. Later this information was kept in bound books and totals were checked against other records.

came from the rolling process, and some of which came from shearing, and then brought them to the office where they were carefully compiled into ledgers. Like the tonnage records, individual records (Figure 10) were compiled into larger, bound ledgers (Figure 11). Lukens attempted to trace the cause of each defect to its source, and these records were essential to that effort. Often disputes arose between the open hearth mills and the plate mills as to who was responsible for specific defects.

The "Defective Plate Record" books, kept from 1894 to 1907, were separated into Open Hearth, Basic Steel, and Acid Steel (the latter two when Lukens installed additional furnaces). Like the Open Hearth Heat books, they were large rectangular books bound in black with gold lettering for the title. They were broken down into month, number of ingots, and the type of defects. In the early version (1895) shown in Figure 11, there were 49 defect categories: narrow end, ragged side & end, cobbled, laminated, sheared wrong, short, burnt, narrow, wide, sand, wrong gauge, blisters, bad surface, pitts & snakes, scale, crooked, wrong size, seams, round end, split end, ragged end, bad bloom, rolled by mistake, and hollow side. On the right side of the page (not shown) there were further defects: buckled, cold pieces, pitts, scabs, hollow end, snakes, layed out wrong, wide & short, ragged side, snakes & scabs, not weight, cinder, snakes and sand, not uniform, graphite, wrong size, housend, caught in guard, pitts & scale, pitts & scabs, bent, split in middle, broke in half, hook rolled on, and caught in transfer.

The process of rolling steel to exact specifications for marine and railroad boilers was still at an early stage, and methods to produce a consistent product had not yet been discovered. In the manufacturing process of rolling steel, "many defects are unavoidable, and that even the most rigid inspection will not suffice to eliminate some defects which are a common annoyance to the manufacturer and customer alike" [6, p. 531]. A good deal of the company's time was spent in trying to locate the causes for these defects. In a note to Charles Lukens Huston, one foreman complained "I find that a good man, working on the Mill Sheets, uses at least half his time recording defectives . . ." [7]. The error recording in these books eventually helped to lead to the answers. As shown in the next chapter, when Charles Lukens Huston began devising experiments and hired a chemist and engineer for testing, they used more intensive writing and communication to find what worked and what didn't. However, prior to that it was still just record keeping.

All of these books were the end product of multiple routines of record keeping by people on the factory floor, and many of them have fingerprints and handprints on them. They are similar to the day book, journal, and ledger in that they record data into a transitional location before entering the final results into ledgers. However, they are different in that they were kept by the foremen on the factory floor. It was a workswide record-keeping project, rather than the work of a single clerk or owner. The processes were becoming more complex, and more people had to participate in the written record keeping in order to gain

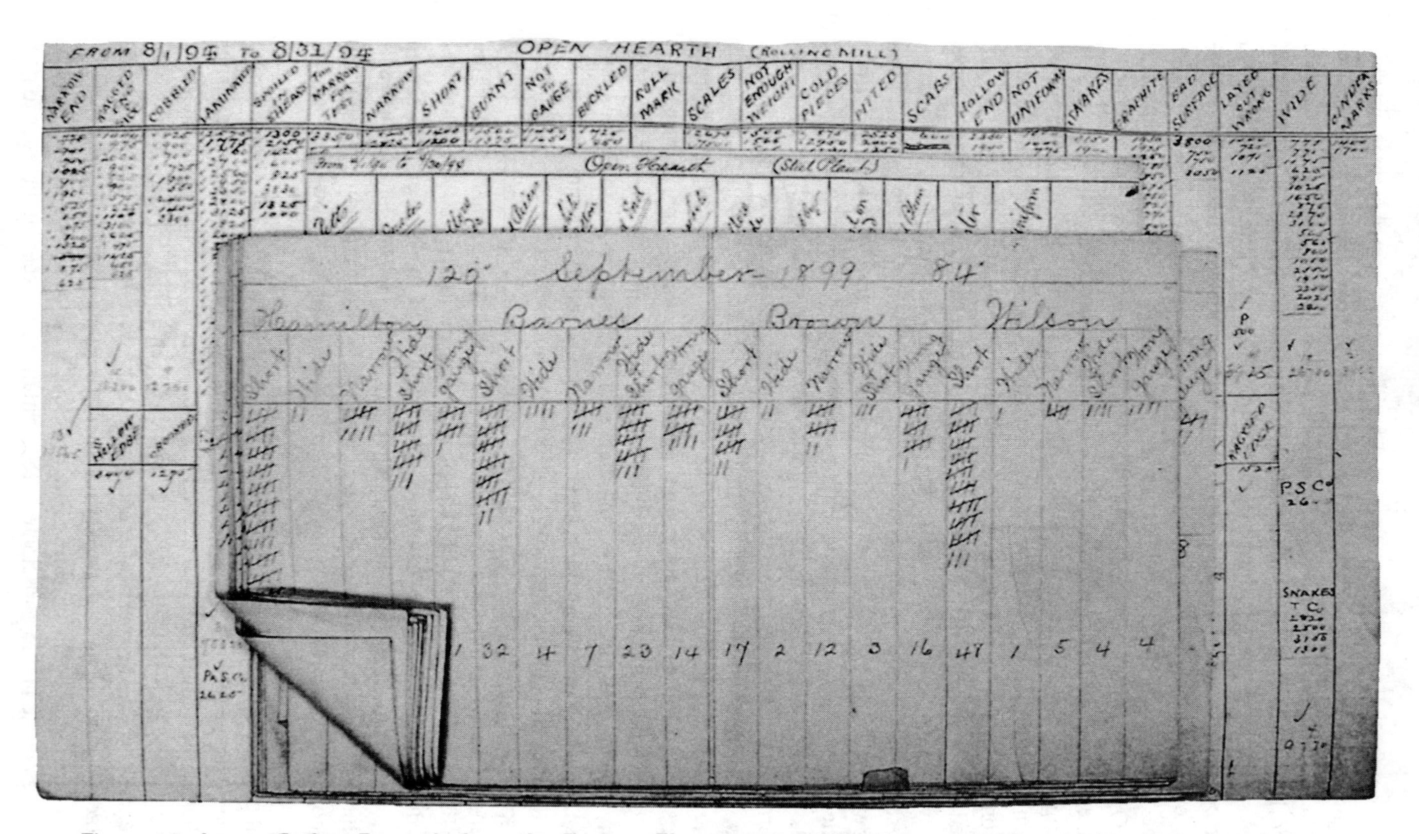

Figure 10. **Loose Defect Records from the Factory Floor (1894-1899).** At the end of the nineteenth century, Lukens began keeping records of the plate defects and their sources. These were later added to a ledger (Figure 11).

FROM AUG 1/94 TO JULY 31/95 — Open

1/94	INGOTS	NARROW END	INGOTS	RAGGED SIDE & END	INGOTS	COBBLED	INGOTS	LAMINATED	INGOTS	SHEARED WRONG	INGOTS	SHORT	INGOTS	BURNT	INGOTS	NARROW
AUG	13	11565	15	16300	9	12700	31 1P	75320 2625	6	10130	12 2P	19280 4360	7	9000	21	36700
SEP	12 2P	15383 3072	5	5810	2 2P	5750 2436	16 2P	45720 2052	9 2P	14566 1958	17 3P	25802 2600	7	7280	9 2P	13675 2078
OCT	5 3P	6135 2465	4 2P	5045 2040	2 2P	2250 2880	10 2	23075 3274 9520	20 2P	40165 3160	5 2P	7370 2302	1 1P	975 1000	3	3875
NOV	4	5755	2	2250	1	1125	4 1P	2580	7	10841	6	5700	1	1200	4	6400
DEC	4	6000	9	9130	10 15	19450 2620	9	26650	3	4000	5	7400	6	6355	10	14010
95 JAN	13	15720	6	6825	4	4625	12	19225	4	4550	31	30340	8	16830	1P 11	5400 12190
FEB	16	17010	6	8935	4	6600	14	26000	6	9950	40	51280	5	5600	15	24180
MAR	3	4825	-	" "	2	2100	18	38895	7	12850	25	31075	1	975	3	3175
APL	11	11405	"	" "	4	3270	24	61890	17	24380	16	16300	3	3925	7	7790
MAY	10	11400	"	" "	4	4135	17	64165	12	14205	8	9870	1	725	17	17925
JUNE	15	21,950	"	" "	4	10250	10	16800	12	33960	14	18700	1	850	12	16986
JULY	12	16610	"	" "	3	2275	20	50200	12	6125	9	8815	5	5020	13	22290
TOTAL	123	150297	49	55835	54	82471	191	466602	120	190479	197	242599	47	60235	128	191653

2/94	INGOTS	WIDE	INGOTS	SAND	INGOTS	WRONG GUAGE	INGOTS	BLISTERS	INGOTS	BAD SURFACE	INGOTS	PITTS & SNAKES	INGOTS	SCALE	INGOTS	CROOKED
AUG	19	26700	8	5816	13	22225	6	14675	4 17	8861 2700	"	" " "	2 17	3473 10173	1	1275
SEP	18 10	38115 1540	2P	4190	6	7850	3	7080	2	3875	4	12325	"	" " "	1 2P	1160 2700
OCT	13	18575	"	" "	"	" "	1	1278	4 1	5526 1329	"	" "	5 2P	11460 8720	3	2285
NOV	4	5570	11 18P	17640 4700	7 3P	9625 4442	2	11640	8 1P	2735 1850	"	" "	2	6800	3	5630
DEC	6	9460	6	14731	17	10990	"	" "	1	800	"	" "	1	1360	"	" "
95 JAN	16	17255	3	58435	11	4925	7	14175	"	" "	7	16226	"	" "	4	2450
FEB	11	13550	18	30025	12	21950	1	2550	7	7900	3	5000	2	4525	5	7760
MAR	10	14625	19	26210	5	7250	2	2200	6	8876	3	5650	2	2175	1	675
APL	11	11445	23	34070	17	25160	9	19915	"	" "	4	9795	2	6350	8	6825
MAY	2	1680	27	44540	12	12575	2	5725	4	4440	2	3100	1	900	6	5875
JUNE	3	5726	11	19925	16	22190	3	6740	4	4275	1	4400	"	" "	"	" "
JULY	6	2055	3	2460	6	11725	4	13506	2	2825	3	5915	11	25670	4	10205
TOTAL	120	165289	129	263245	126	165802	40	99493	40	53483	27	62310	31	75706	38	46820

Figure 11. **Open Hearth Defects Ledger, Upper Left Side (1895).** These are the two top (of three) rows of defects in a 14 by 16-inch ledger. The facing page also has three rows and eight columns of other types of defects.

the knowledge necessary to analyze and control the industrial environment. It wasn't until writing complemented such communication between parties and acted as a reporting device for research, however, that the industrial processes became more predictable. Technical communication was not yet advanced enough to enable people to discover and describe the exact qualities of steel.

Inventories and Payroll Records

Although the inventories and payroll records are interoffice business communication and record keeping, they are interesting in that they, too, reflect the increasing complexity of the works. Prior to the Civil War, when the plant had 17 employees, it was not necessary to list their names. However, by 1900 the number of employees was in the hundreds, so Lukens began collecting their names in a book. The inventories sprang up for different purposes: they were important in making an account of the total property value for taxation. Inventories that included every building shed, tool, and piece of office equipment were time-consuming to create and required an office staff that was diligent and literate. Both the payroll books and the inventories were kept in bound books like the furnace journals.

Payroll records began as loose sheets which, like the tonnage and defect records, originated with the foremen of the different departments. The incoming sheets listed the name of the foremen and the names of the men working under him, but only occasionally made reference to their job functions. Like the car and tonnage records, they were then entered, by a single hand, into a bound journal, which listed names and the rates by which they were paid. The system was loose and troublesome. Employees came and went, and those working on a tonnage basis sometimes argued for changes in the pay when tonnage was sometimes listed differently in different places. There was no systematic management at Lukens Steel yet.

In a separate cost analysis, the basic job positions at the mill were described [8]. They were separated into plate wages, blacksmiths, mill clerks and manager, office clerks and manager, puddle-mill wages and manager, machinists, the firm, furnace mason plate mill, furnace mason puddling mill, car unloading, general labor, stock furnace, flanging machine, and hauling iron. The majority of wages went to plate-mill workers and the second largest to puddle-mill workers. The other largest expenditures were the firm, the office clerks and manager, the mill clerks and manager, and general labor. Between 1880 and 1890 there were approximately 250 to 400 men working daily. By 1891 Lukens had systemized the process to the extent that all the plant employees were listed, biweekly, with their hours and rates. These lists were not alphabetized. The number of men working changed constantly, and no one person knew at any given time how many people were employed at Lukens Iron & Steel Co. As the company grew even larger in the early twentieth century, this became a problem.

EARLY TESTING AND STANDARDS

During the nineteenth century there were many steamboat and railroad engine explosions, which took thousands of lives and made travel dangerous. Much of this is recorded in R. John Brockmann's two books, *Exploding Steamboats, Senate Debates, and Technical Reports* (2002) and *Twisted Rails, Sunken Ships* (2005). There are other books about the issue such as *Death Rode the Rails* by Mark Aldrich. Lukens played an important backstage role in these dramas: first, they experimented continually to try to find out how to make the safest possible product, and also Dr. Huston, and later his son, Charles Lukens Huston, contributed to the social discourse community that created standards.

Lukens was always aware of their limitations. Charles Lukens Huston wrote, "Many of our plates went into the Mississippi River boats and, although I do not know that any boilers made from our plates ever blew up, yet, if they did not, that was due to an overruling Providence rather than anything else, because we had no means of saying exactly what our plates would stand" [9, p. 461]. Prior to 1875, "For the men working at the mill, the science of metallurgy amounted to a series of maxims and a bundle of recollections gathered at painful cost over a period of time. It was all very much on a recipe basis. Every heat amounted to an experiment" [2, p. 42]. In the case of Lukens, they understood that their knowledge was incomplete and worked continually to improve it. Ultimately, being able to track the defects in the steel and trace them to their sources required work that could only take place by using writing and quantitative data analysis as tools.

Steamboat inspection began in 1838, but at that time the inspectors were paid by the vessel owners. Some improvement came with the Steamboat Act of 1852, which provided money for regional inspectors to supervise local inspectors and also made the vessel owners pay licensing fees, rather than paying the inspectors directly. In 1871 Congress passed legislation that created the Office of the Supervising Inspector-General to oversee this department, and it was then that experts began to meet to create standards [10]. Dr. Huston was chosen to chair a committee to set standards for boiler plate in 1877. In 1875 he purchased a testing machine (Figure 12). He made metallurgical experiments, and in 1878 and 1879 he published two articles in the *Journal of the Franklin Institute,* "The Strength and Ductility of Iron and Steel Boiler Plate, at Different Temperatures" and "The Effect of Continued and Progressively Increasing Strain upon Iron." There is no documentation in the archives surrounding these papers; although Huston doubtlessly used paper to create his text and drawings, it was not yet considered important enough to keep.

Dr. Huston's main thesis in his articles was that "the mere question of tensile strength may not be very difficult to ascertain, but the question of '*homogeneousness, toughness, and ability to withstand the effect of repeated heating and cooling,*' is a very different one, and involves a large degree of knowledge in matters still in more or less doubt with the most eminent scientists of the

IMPROVED TESTING MACHINE,

FOR ASCERTAINING THE STRENGTH AND STRETCH OF METALS.

PATENTED AUGUST 6th, 1872.

MANUFACTURED BY

FAIRBANKS & EWING,

PHILADELPHIA.

Fig. 1.

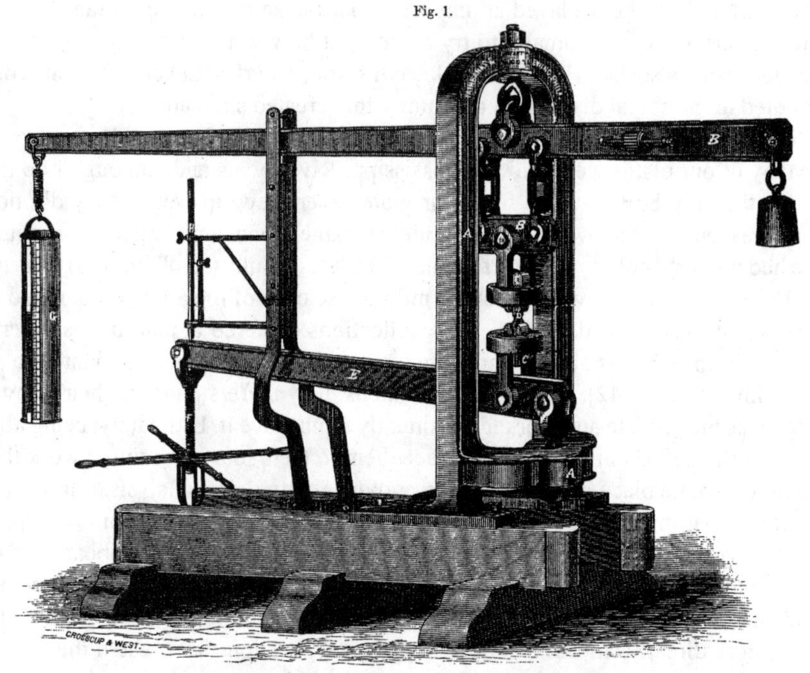

DESCRIPTION OF THE MACHINE.

Fig. 1 represents the machine complete, ready for testing the "tensile strength" and "stretch" of the piece of metal attached.

The cast-iron frame A is in the form of an arch, resting upon timber foundations; within this frame is hung the compound crane-beam B, from which is suspended a wrought-iron socket holding the cast-steel clamps C C, the faces of which are cut file-shaped.

To the fulcrum of the multiplying lever E is attached a similar socket, holding corresponding clamps, C′ C′, pointing upwards.

The piece of metal D, to be tested, is placed within the clamps, where it is held firmly in position. The extreme end of the multiplying lever E is connected with a screw F, which is operated upon by handles revolving on a swivel fastened to the timbers.

To the point pivot of the crane-beam B is suspended the graduated tank G, indicating the weight in pounds, by the use of water.

The attachment H indicates the amount of stretch to the square inch of the metal D before breaking.

Figure 12. **Plate Iron Testing Machine (1873).** Later Dr. Huston endorsed an expanded version of this testing machine with a quote in a Fairbanks & Ewing sales pamphlet.

day" [11, p. 93]. Dr. Huston did not come close to providing answers; he provided questions. His work was then carried forward by Charles Lukens Huston, who also repeatedly stressed that tensile strength alone was not sufficient to judge the quality of steel.

A discourse is a group of statements with underlying assumptions and rules, and a social discourse community is a group formed to discuss a specific subject that adheres to these rules. The discourse community surrounding the creation of standards at first consisted of government officials, professional associations, and manufacturers, but later included scientists, users, and consultants as well. Huston was one of the earliest contributors to the community to set standards for the rolling of boiler plate, but little evidence of his activities exists. He did, however, draw specifications for what he described as "apparatus used for measuring the elasticity and permanent set" that would later be called an extensometer (Figure 13) [12, p. 42].

PATENTS

Although Lukens Steel did not use much written technical communication in the plant until the early twentieth century, they began collecting technical information from the outside world by collecting patents. The earliest patent in their files is from 1838 for "Improvement in the Mode of Smelting Iron Ores." The company collected patents through 1865, mostly relating to the processes of making wrought iron direct from the ore and also improvements in smelting [13]. In 1869, the year before Lukens built the steam-powered mill and puddling furnaces, they collected three patents on improved furnace processes and between

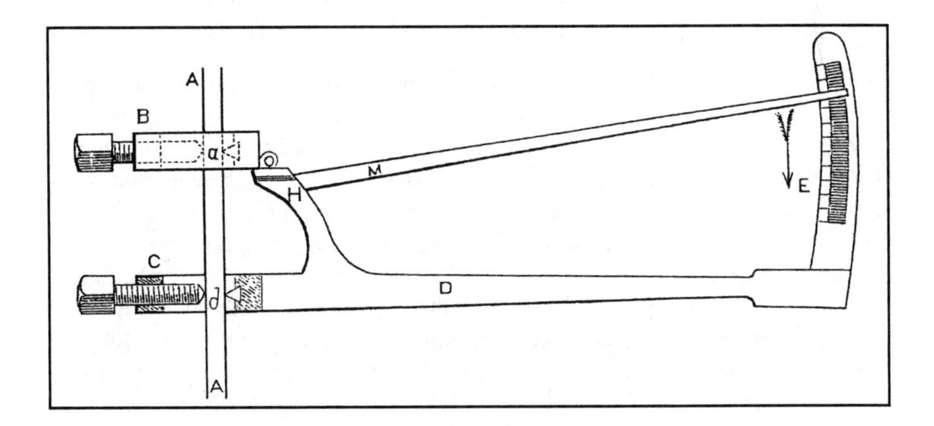

Figure 13. **Dr. Charles Huston's Testing Apparatus (1879).** Dr Huston designed this apparatus, later called an extensometer, to test increasing strain on metal and described it in an article in the *Journal of the Franklin Institute.*

1871 and 1899 they collected 22 more patents. In collecting patents, they were collecting knowledge in the form of technical communication, since the patented processes were explained in writing and illustrated with drawings.

Later, Dr. Huston's son, Charles, joined the ongoing discourse community of patentees. In 1900 he received his first two patents, a "Roll Relieving Device for Rolling Mills" and an "Adjusting Mechanism for Universal Mills" [14]. Probably his most widely adopted innovation was the "Plate Straightening Machine," which he patented in 1902 (Figure 14). He took out three other patents between 1903 and 1920. His final patent, No. 1,440, 221, granted Dec. 26, 1922, was for the supporting roll system that he had built in the four-high mill (Chap. 6, Figure 1).

The patents from the United States Patent Office all followed the same format: they were numbered and dated with full titles and the name and address of the patentee. If there were drawings (and there usually were), they came before the written description. The form and content of the patents were important because they reified a mechanical innovation that was decreed, by an established discourse community, to be significant and new. Thus, it is technical communication as a linguistic "performative" with legal implications. Prior to the twentieth century, technical communication was still the purview of a smaller discourse community—inventors, members of professional associations who wrote and published papers, business owners and their clerks, government officials and engineering consultants. The writing actually used within the plant was still minimal and, in the case of Lukens Steel, limited to recording quantitative data. Technical writing had not yet reached the factory floor.

THE EVOLUTION OF LETTERBOOKS AND THE INTRODUCTION OF THE TYPEWRITER

During the years 1870 to 1900 the letterpress books increased from one or two per year to ten or more. Moreover, in 1888 they were joined by telegram press books, which held two outgoing telegrams per page and increased in frequency from one or two books per year to six per year in 1903. These letterbooks and telegram books became a physically massive recording project, and constricted communications because each letter or telegram had to go to a central location, while the ink was still wet, to be pressed onto a damp page. As the plant grew, it became more decentralized. In 1903 the reign of the letterbook was over: managers and executives began keeping individual files. From that point forward, however, we have only the files of the works manager, Charles Lukens Huston. Still, as works manager, his viewpoint is a good one from which to view the operation of the plant and the use of the paper therein.

The contents of the letterpress books show an increase in the variety of contents over time. For instance, the firm had stamps made that created templates for repetitive tasks. The template edge was usually in a different color (pink, purple,

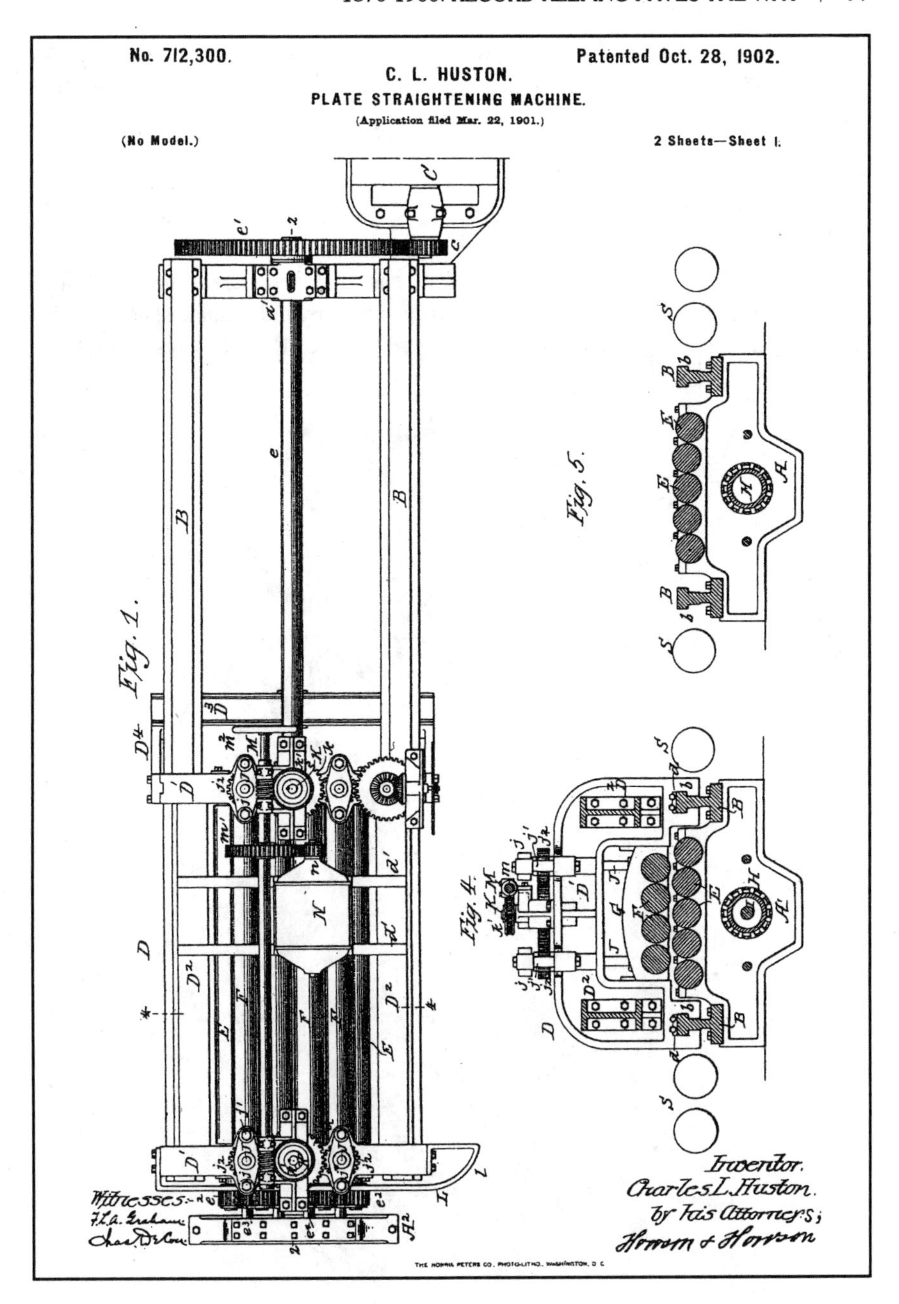

Figure 14. **First Page of Charles Lukens Huston's Plate-Straightening Patent (1902).** In United States patents, drawings came first and then the text that described the drawing.

or blue), and the writing within it was standard black ink. These stamps were for check receipts, price lists, and incoming and outgoing order forms. The earliest stamp, in 1879, was a template for a price list (Figure 15).

As the office staff grew, so did the number of different hands writing in the letterbooks. Until 1890, most of the letters were signed with the firm's business name, which reflected the (shifting) existing partnerships: Gibbons & Huston, Huston & Penrose, Huston, Penrose & Co., and finally Chas. Huston & Sons. In 1890 the firm incorporated as the "Lukens Iron and Steel Company" and, for the first time, the letters were consistently signed by individuals. The most frequent signatures in the letterpress books were those of R. B. Haines (Secretary),

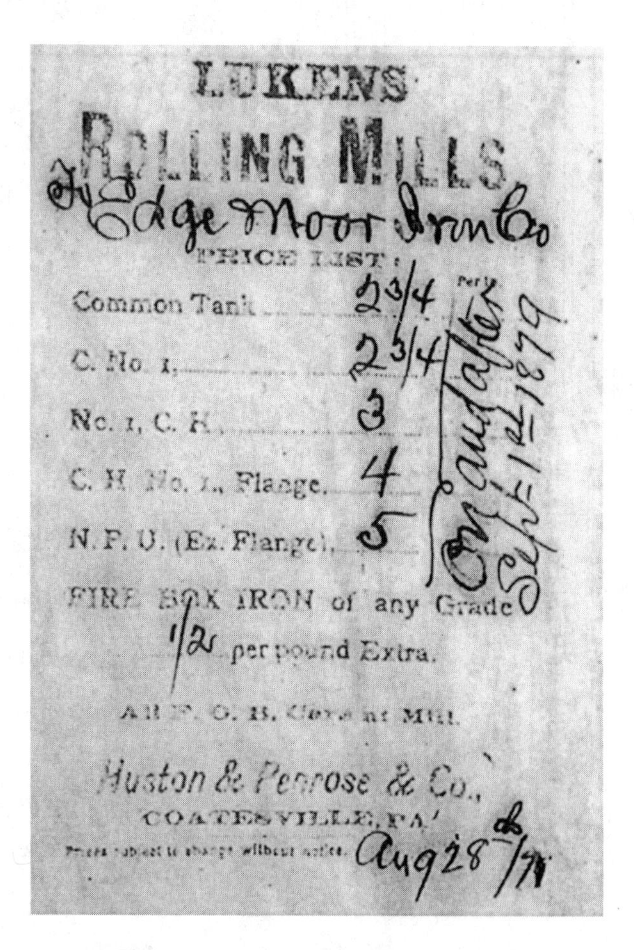

LUKENS
ROLLING MILLS
To Edge Moor Iron Co
PRICE LIST:
Common Tank 2 3/4 per lb
C. No. 1, 2 3/4
No. 1, C. H. 3
C. H. No. 1, Flange 4
N. F. U. (Ex. Flange), 5
On and after Sept 1st 1879
FIRE BOX IRON of any Grade
1/2c per pound Extra.
All F. O. B. Cars at Mill.
Huston & Penrose & Co.,
COATESVILLE, PA.
Prices subject to change without notice.
Aug 28th/79

Figure 15. **Specialized Stamp (1879).** Stamps such as this one began appearing in the letterpress books in the 1870s. This stamp was small enough that four could fit on a page.

A. F. Huston (Vice President, then President), J. R. Van Ormer (General Sales Ac't.), Jos. Humpton (Treasurer), F. H. Gordon (Sales Mgr.), Charles Lukens Huston (Gen'l Mgr, then Vice President), and H. B. Spackman (Purchasing Agent). Other unidentified people filled out standardized stamped template forms as well. The Lukens Iron and Steel office had become an organized system administered by multiple people (Figure 16).

Still, the main content of the letters were shipping and billing instructions, sales negotiations, and logistical communications for the movement of product, rather than technical communication. The letters occasionally contained specifications for the sizes and shapes of plate being ordered. Often the plate specifications were just lists of sizes, but sometimes they were accompanied by handmade drawings. Also, if a plate failed, the technical reasons for the failure were discussed and conclusions communicated by mail. The letterbooks were important, because they established a legal record of transactions [15]. The letterpress books were indexed, and after 1890 the individual entries were checked off against the index in red pencil. It was a serious and cumbersome method of record keeping.

The greatest leap forward for technical communication during this time was the adoption of the typewriter. The typewriter was introduced into the letterpress books in a letter typed by a Remington salesman on June 7, 1885 (Figure 17). Gradually, typed letters replaced handwritten letters in the letterbooks. This allowed more technical communication to occur on paper, in writing, than before because typing was clearer and easier to read. Typing made accurate and rapid written communication possible.

Extensive technical communication started when Charles Lukens Huston began building the 120-inch mill in 1889. It was the largest mill in the United States for a long time, and the details of the designing and construction were worked out in letters going back and forth between Huston and the makers of the parts (rolls, housing, building, etc.). Written prose was sufficient to plan, design, and convey meaning, although drawings were sometimes required as well. For instance, a letter dated June 1, 1887, requested additional drawings: "Please send us another sketch for the former, as it could be made a dozen different ways from the present sketch and still conform to instructions."

In the case of Charles Lukens Huston, typewritten letters were especially important for ease in conveying meaning. His handwriting was difficult to read (Figure 18). Typing removed that barrier (Figure 19). Thus, he was able to brainstorm with various firms producing portions of the machinery to choose the best possible solution for each component of the new mill. The back-and-forth dialogues between Huston and the suppliers were frequent and continued until the mill was opened. The following extended quote of a letter dated February 11, 1889, shows the amount of detailed, complex technical communication that can be communicated through prose:

Figure 16. **Lukens Iron and Steel Office (1895).** Left to right: H. B. Spackman, Irwin Vannan, Mrs. Anna E. Martin Ford, Joseph Christie, Dr. Charles Huston, Frank Russell, Joseph Humpton, May Humpton, Howard Humpton, Mr. A. F. Huston, Anna Lawrence, Owen Spackman, Mr. Clayton Emery, Charles F. Humpton.

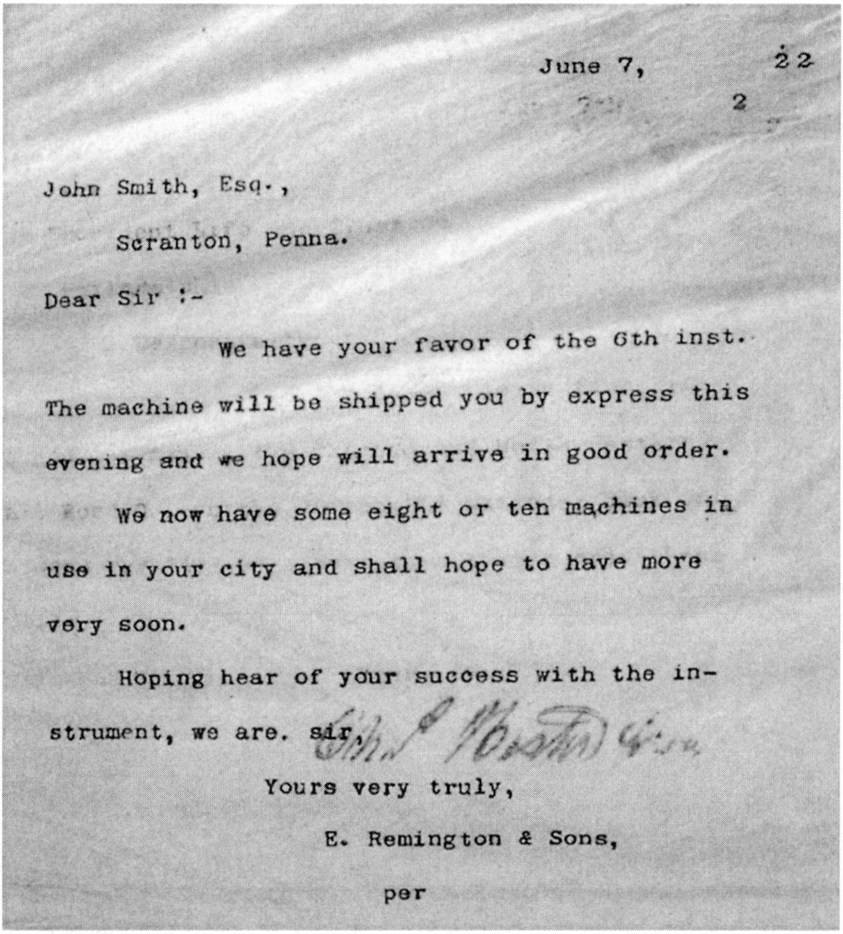

June 7, 22

2

John Smith, Esq.,

Scranton, Penna.

Dear Sir :-

We have your favor of the 6th inst. The machine will be shipped you by express this evening and we hope will arrive in good order.

We now have some eight or ten machines in use in your city and shall hope to have more very soon.

Hoping hear of your success with the instrument, we are, sir,

Yours very truly,

E. Remington & Sons,

per

Figure 17. **First Typewritten Letter in Lukens Letterpress Book (1885).** This letter was a demonstration, by the Remington salesman, of how to type and then press a letter into the book. The signature on the following page can be seen through the tissue-thin paper.

In regard to the roll train, we do not recollect clearly whether you arranged to give the additional height to the housings to maintain the same slack-up that we would have had with 32 inch rolls or not, but trust you will not lose sight of this point. We understand that 16 inches was what the Park mill had. In regard to the amount of slack-up, the Otis people in their new mill have provided for two feet, so that we should incline to think 16 inches would be little enough for us to start out with, although we may not have occasion to use more then 10 or 15 inches for some time to come [16].

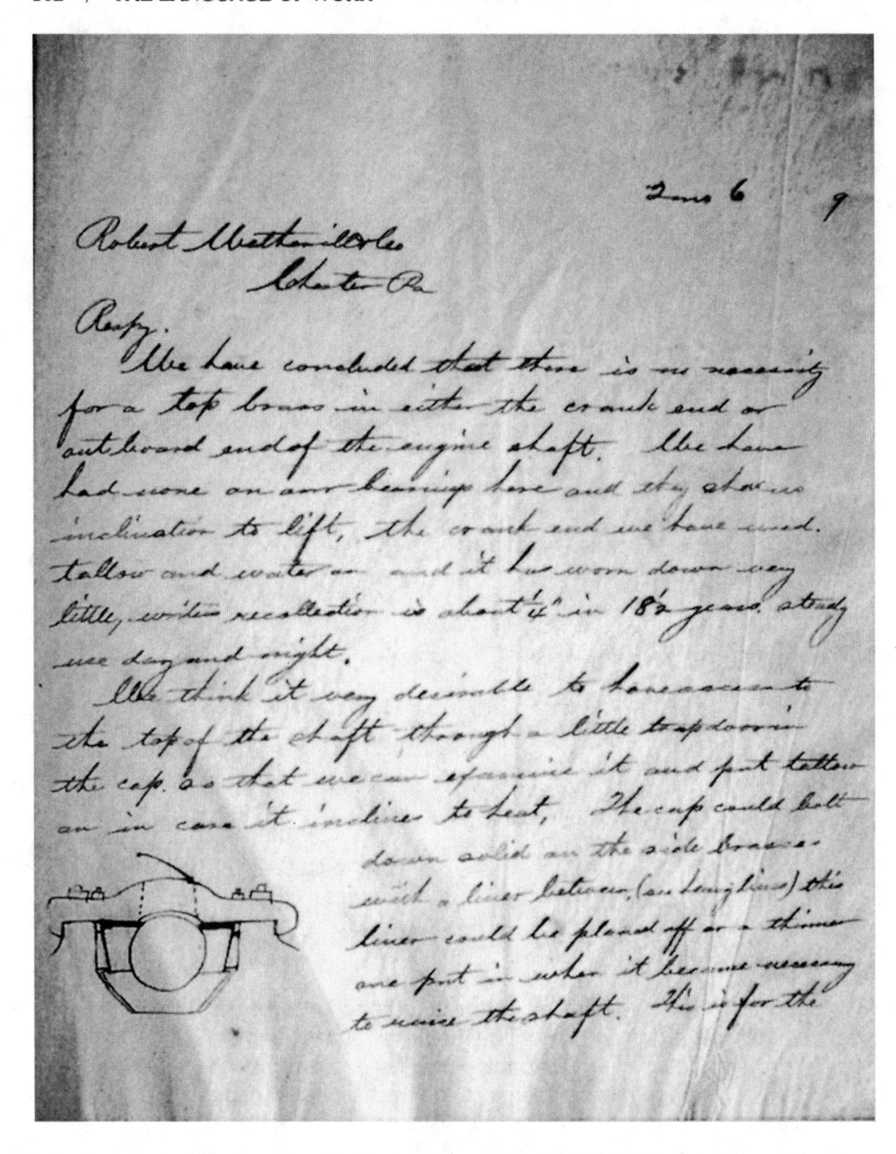

June 6 9

Robert Wetherill & Co
Chester Pa

Resp'y.

We have concluded that there is no necessity for a top brass in either the crank end or outboard end of the engine shaft. We have had none on our bearings here and they show no inclination to lift, the crank end we have used tallow and water on and it has worn down very little, writer's recollection is about ¼" in 18½ years steady use day and night.

We think it very desirable to have access to the top of the shaft through a little trap door in the cap so that we can examine it and put tallow on in case it inclines to heat. The cap could bolt down solid on the side brasses with a liner between (as long liner) this liner could be planed off or a thinner one put in when it became necessary to raise the shaft. This is for the

Figure 18. **Handwritten Technical Communication in Letterpress Book (1889).** This letter by Charles Lukens Huston regarding the planning and construction of the 120-inch mill, is an example of his handwriting, which could be difficult to read.

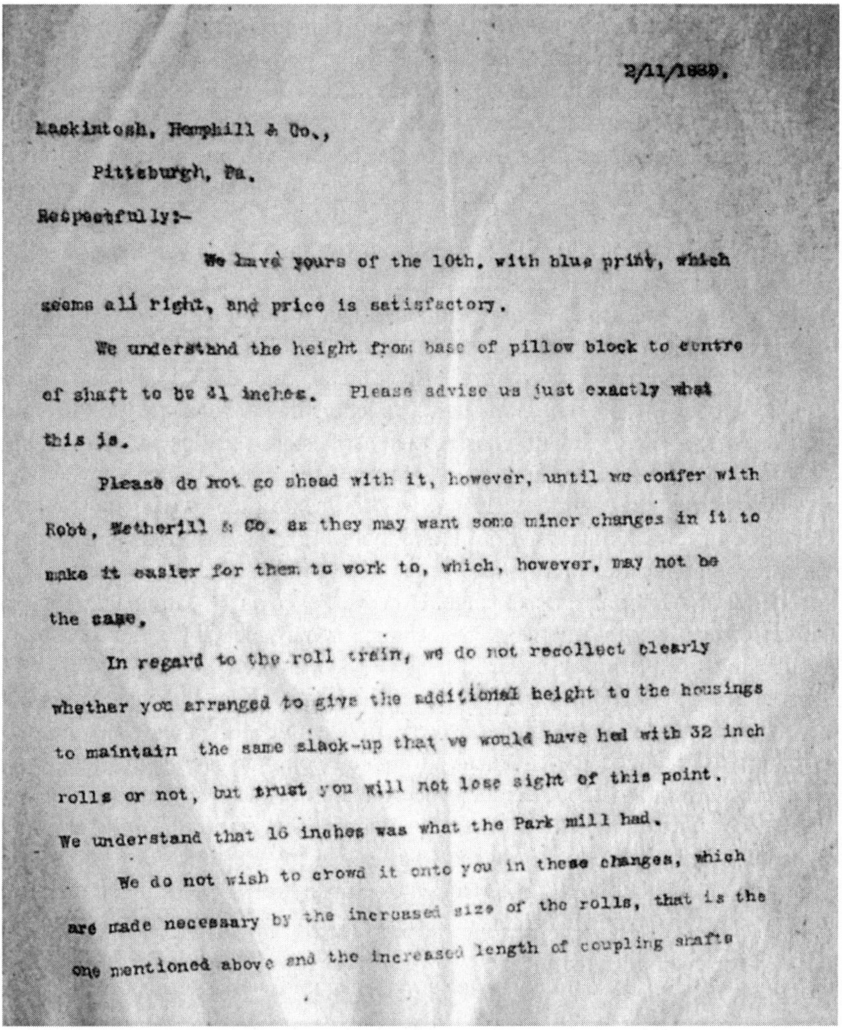

2/11/1889.

Mackintosh, Hemphill & Co.,

Pittsburgh, Pa.

Respectfully:-

We have yours of the 10th. with blue print, which seems all right, and price is satisfactory.

We understand the height from base of pillow block to centre of shaft to be 41 inches. Please advise us just exactly what this is.

Please do not go ahead with it, however, until we confer with Robt, Wetherill & Co. as they may want some minor changes in it to make it easier for them to work to, which, however, may not be the same,

In regard to the roll train, we do not recollect clearly whether you arranged to give the additional height to the housings to maintain the same slack-up that we would have had with 32 inch rolls or not, but trust you will not lose sight of this point. We understand that 16 inches was what the Park mill had.

We do not wish to crowd it onto you in these changes, which are made necessary by the increased size of the rolls, that is the one mentioned above and the increased length of coupling shafts

Figure 19. **Typewritten Technical Communication in Letterpress Book (1889).** This letter, also by Huston on the same topic, demonstrates that typing aided legibility. Typing made accurate and rapid written communication possible.

In this excerpt we can see that technical communication is not the mere transmission of known data. Like Foucault suggested, there was a dialogue that went before and a dialogue that would come after. The author is taking part in an asynchronous conversation, trying to recall past parts of the conversation, stating some requirements ("addition height to the housings"), and referring to other participants in the ongoing conversation ("the Park mill" and "the Otis people"). It is an artifact of an ongoing social discourse community, a web that would gradually create a rolling mill. It is also an example of how writing aids the thought process and the negotiation of ideas. When Huston wrote "we should incline to think 16 inches would be little enough," that idea is the end product of multiple comparisons with the rest of the discourse community.

In 1903 letterpress books were discontinued, and each individual began keeping his or her own system of files. The letters that are analyzed in this book are from the point of view of Charles Lukens Huston, since he saved them and later his family deposited them at the Hagley Museum and Archives. As vice president and works manager, Huston was the central contact for most events in the open hearth furnaces and mill, and therefore, from his viewpoint, we can see the communication that spread out into the mill and created networks of knowledge and information exchange. However, even as we can imagine him as the center of a growing network of technical communication, we also need to imagine the amanuensis at his side.

From the introduction of the typewriter forward, Charles Lukens Huston had a stenographer and typist with him very frequently. His letters (and those of many others) retained the cadences of spoken communication, but they were impeccably typed, without errors or corrections. Thus, we can presume the emergence of a new type of worker: the stenographer typist. There is only one slim file of correspondence between Charles Lukens Huston and his secretary, Miss Robertson, after he resigned from the company in 1925. Until that time, she took his dictation, typed his letters, and filed his papers. Without the assistance of this new type of literate and detail-oriented worker, communicating complex technology by prose would not have evolved as rapidly or as efficiently.

* * * * *

Lukens' stance of "riskless sufficiency," which took them through the difficult years of 1800 to 1870, was abandoned in 1870 when they built their first steam-powered mill. From that point forward, new forms of record keeping and communications evolved with the rapidly growing company. In the years between 1870 and 1890 this mainly consisted of several systems of ledgers, some of which evolved from the practice of keeping accounting journals: there were puddle-mill journals, tonnage books, and, most importantly, detailed lists of plate defects. Many of these records originated on the factory floor as workers kept data in their own notebooks or reams of paper. Due to the increasing push toward safety, Dr. Huston became involved in the creation and negotiation of boiler plate standards and published two articles on the strength of iron plate under different conditions. However, the most important advance for the future of technical communication during this period was the adoption of the typewriter in 1885 and the emergence of the stenographer typist, who could change spoken words of technical experts into readable prose. Once this had become possible—that technical experts in a variety of specialties could communicate important details about the behaviors of materials across space and time in an accurate and predictable way—the stage was set for increased communication, increasingly perfected technology, and an increased ability to collaborate to create standards. At Lukens Steel, technical communication became a companywide activity in the twentieth century.

REFERENCES

1. G. G. Eggert, *The Iron Industry in Pennsylvania*, The Pennsylvania Historical Association, University Park, Pennsylvania, 1994.
2. J. C. Skaggs, *Lukens, 1850-1870: A Case Study in the Mid-Nineteenth Century American Iron Industry*, dissertation, University of Delaware, Wilmington, Delaware, 1975.
3. *206" Mill, 1916-1958, Lukens Steel Archives*, Hagley Museum and Library, Wilmington, Delaware, 1922.
4. A. Holly, The Field of Mechanical Engineering, *Transactions of the American Society of Mechanical Engineers,* 1882.
5. J. Yates, *Control Through Communication: The Rise of System in American Management*, Johns Hopkins University Press, Baltimore, Maryland, 1989.
6. J. M. Camp and C. B. Francis, *The Making, Shaping and Treating of Steel* (4th Edition), Carnegie Steel Bureau of Instruction, Pittsburgh, Pennsylvania, 1924.
7. "Reports & Memos from Dep't Heads & Workmen," B-4, *Lukens Steel Archives*, Hagley Museum and Library, Wilmington, Delaware, 1900-1914.
8. "Cost Accounting Charts for Steel Plant & Mills," B-183, *Lukens Steel Archives*, Hagley Museum and Library, Wilmington, Delaware, 1903.
9. C. L. Huston, *132 Years without Losing a Customer, Systems, the Magazine of Business*, pp. 461-469, April 1925.
10. L. M. Short, *Steamboat-Inspection Service Its History, Activities and Organizations*, D. Appleton and Company, New York, 1922.
11. C. Huston, The Strength and Ductility of Iron and Steel Boiler Plate, at Different Temperatures, *Journal of the Franklin Institute, 105*:2, pp. 93-97, 1878.
12. C. Huston, The Effect of Continued and Progressively Increasing Strain Upon Iron, *Journal of the Franklin Institute, 107*:1, pp. 41-44, 1879.
13. "Non-Lukens Patents on Iron Manufacturing," B-2216, *Lukens Steel Archives*, Hagley Museum and Library, Wilmington, Delaware, 1838-1899.
14. "Patents issued to C. L. Huston," B-17, *Lukens Steel Archives*, Hagley Museum and Library, Wilmington, Delaware, 1900-1922.
15. M. Hodgson, *Conservation Distlist—Letterbooks*, accessed Sept. 8. 2007, from http://palimpsest.stanford.edu/byform/mailing-lists/cdl/2004/1259.html, 2004.
16. C. L. Huston, "Letter to Mackintosh, Hemphill & Co," B-1245, *Lukens Steel Archives*, Hagley Museum and Library, Wilmington, Delaware, 1889.

CHAPTER 5

Lukens 1900-1915: An Explosion of Technical Communication

The social discourse community at Lukens Steel exploded from comparative silence into a multiplicity of voices as operating the plant became more complex and technical communication became a necessary component of almost every process. The foremen of each unit had to be able to communicate to the managers of each division who, in turn, had to communicate with one another in addition to the people who did the testing and Charles Lukens Huston himself. Most of the types of communication described in the previous chapters—data collection, record keeping, correspondence, etc.—continued on, but there were also board meeting minutes, handwritten notes between Charles Lukens Huston and the department heads, drawings and blueprints, boiler testing documents, published articles in trade journals, product descriptions, and testing documents (letters, drafts, meeting minutes, transcriptions of arguments, papers, and publications). This was no longer solely the voice of the owner or the partners; many people in the plant participated, and this is where technical communication emerged and grew.

Work on improving the plant never stopped. Interplant communications about construction became essential. This communication took place in writing and drawing. Technical communication had become a method of working out problems by combining multiple opinions and ideas. In some construction projects, writing was the major means of communication: when building the "big mill," the majority of the interaction between Lukens and United Engineering & Foundry took place by letter. Workers, foremen, and managers used drawings and blueprints, which had become crucial to any design or construction process. The documentation of testing exploded. There were at least four forms of testing at Lukens Iron and Steel Co.: testing as part of the manufacturing process; in-house testing of plant equipment; scientific testing for the purposes of discovery and publication; and Charles Lukens Huston's personal interaction with the social discourse community that created national boiler plate standards. At this time Lukens also began publishing articles, advertisements, and product guides as a form of public relations. As a consequence of this sudden increase in technical communication, the 15 years covered in this chapter will take as long to analyze as the first 90 years described in the previous chapters.

THE TECHNOLOGICAL PROCESS: 1900 to 1915

At the turn of the nineteenth century, Lukens Steel embarked on an expansion plan. In 1899 they built four more open hearth furnaces on the other side of the Brandywine River next to a new 48-inch three-high universal mill that could roll the edges of a plate as well as the sides. In 1903 they built a three-high 140-inch mill in a separate plant with furnaces, cranes, shears, leveling rolls, shipping facilities, and automatic tables. Like the 120-inch mill, at the time it was built, it was the largest rolling mill in the United States. Because these mills were unusual, their planning and construction required a great deal of communication between Lukens and the suppliers of the parts. The older two-high mill had rolls that moved in opposite directions so that the motion of the rolls helped to pull the iron or steel through. A three-high mill has three rolls so that the plate can pass through the rolls in both directions without reversing the movement of the rolls or handing the plate back from the catchers to the feeder's side [1, p. 683]. The universal mill had, in addition to the horizontal rolls, vertical rolls that finished the edges of plate metal [2, p. 446]. The final layout of the mills was negotiated, planned, and revised in letters and drawings. Technical communication is not mere documentation of existing technology, it is an essential part in the process of building it.

Although this book separates the chapters by time, in reality the technological changes at Lukens were seamless. After the relatively stagnant period described by Skaggs as "riskless sufficiency," Lukens fixed parts of the plant and added other components continually. The constant changes in the layout of the plant and in lifting, storage, shipping, and energy systems are documented in blueprints—the machinery was in a constant state of flux. The 1890 120-inch mill was rebuilt into a 134-inch mill in 1900 and reduced to a 112-inch mill in 1903, when the new 140-inch mill was finished [3, p. 3]. The heating processes were continually modified (especially when they didn't work). The 140-inch mill was built with two soaking pit furnaces and three continuous furnaces in an adjacent structure to heat the slabs and ingots before rolling (Figure 1). Eventually they found that the continuous furnaces were not working well so they were replaced by two additional soaking pits [4]. Moreover, the supporting structures such as steam engines, gas producers, cranes, drainage, tracks for the railroad cars that moved the steel, testing facilities, and offices were continually updated as well. Lukens Steel was a work in progress, which was now dependent on continual discussion through technical communication.

Figure 1. **Reheating Furnace.** Ingots or slabs of steel that did not go directly from the open hearth soaking pits to the rolls had to be heated in separate reheating furnaces.

Although Lukens did not have a master list of workers in the plant, Camp and Francis describe the "mill force" (the people that were necessary to run the plant). The following extended quote gives an overview of the personnel structure at a rolling mill:

> Under the general superintendent of the steel plant there may be a number of rolling mill superintendents, each of whom will have charge of a group of mills turning out similar products. As his assistants, the mill superintendent selects foremen, each of whom are responsible for the successful operation of one or two of the mills. Below the foreman, the mill is divided into departments, with a man at the head of each, who is charged with the performance of a certain part of the work. Thus, there is the heater who has the heating of the material to look after; the roller, who superintends the actual rolling process; the engineer who tends the engine, or an electrician, if motors are used for running the mill; and the shearmen, whose duty is to see that the product is properly cut. Besides these, other departments, such as the machine and the electric shop, the inspection and shipping departments, play important parts in the mill operation, though they do not come under the direct authority of the mill superintendent. When it is remembered that the failure of any one of these may close down the whole mill, the importance of system and of the personnel of the organization is more fully appreciated [2, p. 452].

Charles Lukens, however, still ran the business as an extended family. Huston was the vice president and works manager, so ultimately all of the plant employees reported to him. Horace Spackman, who had joined the company as an office boy in 1881, became the purchasing agent, then a director, and eventually the second vice president in 1900 [3, p. 136]. Thus, he was almost as important as Huston in the running of the plant. Other long-time employees, such as William Hamilton, the manager of operations, P. C. Haldeman, the master mechanic, and A. Goodfellow, mechanical engineer and draftsman, had powerful voices in running the plant and communicated with each other continually in writing. Although they sometimes held official job titles, they all did a wide variety of work across divisions. These are only some of the voices that are raised, like a chorus, in the early twentieth century.

Just as the plant itself was in continual development, the industrial systems within it were constantly being modified and perfected. Adding new buildings and machinery, seeing if they worked, and correcting if they did not were major sources of the explosion of technical communication. There were many factors that had to be discussed, solutions proposed and tried. The discussions took place in person but also in writing and drawing. Rolling mills were complex:

> In the rolling of steel there are five factors to be considered, namely, the temperature of the steel during the rolling, the chemical composition of the metal, the speed at which the rolls are revolved, the draught in each pass, and the diameter of the rolls. Furthermore, these factors should be considered from the three different standpoints of power, or energy, required to deform the steel; their effect upon the rolling properties of the metal, that is, the way it will spread, bend and flow in the rolls; and their effect on the quality of the finished product. All these matters have not been fully investigated and our knowledge concerning them is somewhat meager [2, p. 453].

Thus, the producers of rolled iron and steel did not know exactly how to do it. They did not fully understand the physics, chemistry, or the engineering, and therefore they were continually communicating about testing, results, successes, and failures. Detailed technical communication was essential to the success of the plant.

The open-hearth process was complex as well (Figures 2 and 3). In order to make ingots of the correct size and shape for each of the mills, pig iron, scrap metal, and chemicals were melted in the furnace then poured into molds for different-shaped ingots. The condition of the mold (hot, cold, surface pits, hairline cracks) had a great deal to do with the resulting ingot as did the way it was poured; for instance, splashing would make scabs of metal on the side surface. Also, the way that the ingots cooled changed their chemical makeup and thence their rolling properties. The best way to cool them was slowly, in a "soaking pit," which was a specially constructed furnace that kept the exteriors of

Figure 2. **Tapping an Open Hearth Furnace.** The molds for filling are in a pit near the furnace so that the steel can be poured directly into them.

the ingots warm while the interior cooled. Whether an ingot was then directly moved to the mill for rolling also made a difference; if it was not, it had to be reheated in a different reheating furnace (one that warmed gradually from the outside in) before being sent through the rolls. The making of ingots was continually negotiated between many people at Lukens Steel; even workers had a voice in the different experiments that they conducted to make their processes work best. Writing about testing was not only essential, but voluminous as they sought ways to solve complex problems. Technical communication had moved into the mainstream of the industrial process, not as a method of documenting the known, but as a method of discovering the unknown.

INTRAPLANT COMMUNICATIONS

In 1903 each manager began to file their own letters (or had a clerk who did so). The unwieldy letterbooks, besides being difficult to read, had to be processed at the same location, so every writer had to bring his daily correspondence to a central office. However, carbon paper was not widely used until about 1912 [5, p. 48]. Thus, an intermediary genre of written communication emerged between 1904 and 1911 between Charles Lukens Huston and the managers and

Figure 3. **Open Hearth Furnaces, Ingot Molds, the Soaking Pit and Crane.** Lukens built open hearths in sets—Open Hearth #1 (1890) had two furnaces, Open Hearth #2 (1899) had four furnaces, and Open Hearth #3 (1918) had eight furnaces.

the foremen. In a folder called "Reports & Memos from Dep't Heads & Workmen" [6] there are hundreds of notes, mostly on 6 × 9-inch scraps of paper, sent back and forth between Huston, his managers, and workers, from various parts of the plants. These notes were usually folded in quarters, with the name of the recipient written on the outside, and traveled through an intraplant mail system. This handwritten method of intraplant technical communication lasted from 1904, when they discontinued the letterbooks, until 1911, when carbon paper was used and the number of stenographer typists increased. After 1911 each manager could dictate rather than write letters, and the length of the letters increased. Moreover, the multiple copies, made with carbon paper, could be addressed to multiple receivers; the network of communications increased in its scope.

The "Reports & Memos from Dep't Heads & Workmen" are often undated notes, letters, calculations, test results, and drawings on a variety of topics that were mainly technical in nature. In these notes we can also see the emergence of the use of printed forms to systemize the information that flowed throughout the works, since various forms are often attached to the notes. Huston communicated most frequently with William H. Hamilton (then called the general superintendent), Alfred Goodfellow (mechanical engineer in charge of plant construction and drawings), F. H. Woodhull (production engineer), H. G. Martin (chemist), Howard Taggert (testing department), and H. C. Moyers. Although these notes are informal, the amount of data that they contained, in test results, drawings, and complex analysis of problems, is impressive. These small notes themselves draw a portrait of the works from 1904 to 1911. Although there are hundreds of them, only two examples are described below: one regarding a failed inspection and another about a mechanical problem. They are an example of a single line of communication, in contrast to the later networks made possible by carbon copies.

One of the slips of paper in this series is a test report with a note written on it from Charles Lukens Huston to Howard Taggert who was the Engineer of Tests (Figure 4). Taggert coordinated inspection from outside inspectors (usually railroad or government), conducted tensile-strength and bend tests, and kept reports of the results. The testing department was a major stop (and bottleneck) within the manufacturing process since railroad cars with unshipped plate could block other shipments. Much of the testing occurred as the product was going out the door. If the plates did not pass inspection, they were rejected, which happened frequently. Rejections were an important generator of technical, mechanical, and chemical research, resulting in increased communication as Lukens tried to find the causes and solve the problems.

The rejection slip, dated 10/30/06, had a note written on it from Huston to Taggert, asking "What were the bending requirements on this job," dated 12/2. Taggert provided a handwritten answer describing why the plates were rejected. First he gave the particulars of the requirements. He wrote:

> Requirements for this order are those of Class "B"—U. S. N. viz 60 m T. S. + 25% Elong in 8" with transverse load bends flat.
> It is Inspector's practice to alternate the tests between "Top" + "Bottom" –
> Four of the bends as noted on report, were taken from the "Top" + as plates did not have sufficient discard they failed –
> On the fifth plate bends were taken from "Bottom" the first bend failed but duplicate withstood the test—the Inspector however was unwilling to accept plates
> 12/4/06 H. T. [6]

This was a fairly standard exchange as they tried to track the complex causes of rejections and defects, and it took place by passing folded notes. They frequently had trouble with the U.S. Navy Class B plates that had to conform to specifications that were hard to achieve and struggled with this problem for years. Writing made knowledge exchange across distances possible.

Another example of the type of communication that was passed by note between Charles Lukens Huston and the foremen and workers was about a piece

LUKENS IRON AND STEEL CO.

TESTING DEPARTMENT

#371

Report of Plates Rejected and to be Replaced. Date, 10 - 30 - 1906

ORDER No.	SLAB No.	MELT No.	No. Pls.	SIZE	T. S.	CAUSE OF REJECTION
				Rej. by Capt. Bartlett.		
1256	50055J	9253	1	130" dia. x 3/4	64800	Failure in bending test
"	507013	14053	1	"	61580	"
"	503055	9253	1-	"	60080	"
"	50305C	9253	1	"	66980	"
"	50350G	9253	1-	"	63840	"

R. C. Hoffman & Co.

Figure 4. **Rejection Report (1906).** Rejection reports were one of the most important standardized forms that were used in the manufacturing process. This one is addressed to Charles Lukens Huston and has a note from him to Howard Taggert so it circulated through the plant, arriving back at Huston's files.

of machinery that had broken. H. C. Moyers sent the following drawing with an introductory note that read "140 Mill Engine Free Exhaust Valve blew Wide open this morning while running Condensing, caused by too strong exhaust" (Figure 5). On a page after the drawing, Moyers concluded, "I would suggest a steam Gate or Electric Value operated in Main Steam Pipe, by a Speed limit on Shaft" [6]. In this note, the author explained the problem, provided evidence of his solution (the drawing), and made recommendations as to how to fix the problem permanently. The act of writing itself helps the author to make his tacit knowledge explicit, and then it can be conveyed to others.

Overall, the "Reports & Memos from Dep't Heads & Workmen" have an immediacy that allows one to hear actual exchanges and troubleshooting among management and workmen. Some of the notes that originated on the factory floor are black with dirt, a testament to the environment in which they were created. These informal notes, frequently supported by test data or drawings, show people interacting to solve problems, implement solutions and, in general, make tacit knowledge explicit so that it can be shared on a broader scale. The "Reports & Memos from Dep't Heads & Workmen" are an informal communication system, closely tied to the factory floor. They are the precursor to the more formal

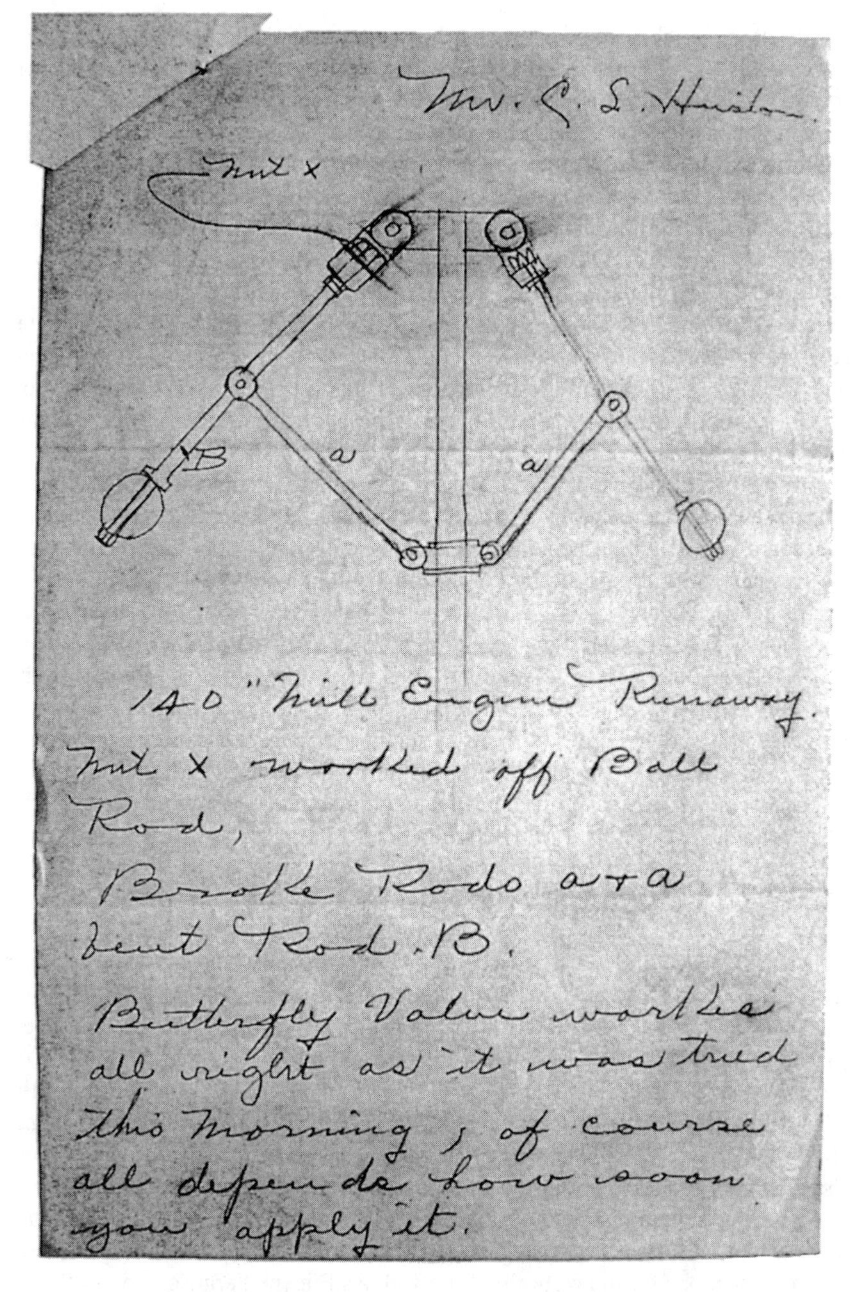

Mr. C. L. Huston

140" Mill Engine Runaway.
Nut X worked off Ball
Rod,
Broke Rods a + a
bent Rod. B.
Butterfly Valve works
all right as it was tried
this morning; of course
all depends how soon
you apply it.

Figure 5. **Note from H. C. Moyer to Charles Lukens Huston.** This note was about a minor breakdown in the plant and described a temporary solution as well as a long-term one.

typed letters. At this point written technical communication was alive but not systemized. The later systemization, with stenographers and multiple carbon copies, evolved from this elemental need that sprang from the factory floor as keeping the machinery working became more complex.

In this group of notes we can see that forms for repeated tasks were beginning to develop. These forms could be ordered from printmakers. For instance, there are several forms in the testing department, some for the laboratory and some for the factory floor. The "Report of Test of Steel" (shown later, in Figure 20) is the largest. The others, like the "Report of Plates Rejected" (shown earlier, in Figure 4), are smaller. There is also a general "Laboratory Report" with the date; Heat Analysis with the number, amounts of carbon, magnesium, sulphur and phosphorous; and a place for the chemist's signature. Technical communication was systemized, first, for record-keeping purposes, as described in the last chapter. The systemization in writing did not come until later.

Efforts toward systemization of writing were made in some parts of the plant, but they were still premature. In about 1900 Charles Lukens Huston instituted a procedure for creating machine shop cards. Before any shop work could start, requisitions had to be issued by the machine shop clerk, who had to record the details before the "work card" was given to a machinist (Figure 6). The card then accompanied the job through the entire process, and the number of hours for different types of work were recorded on it. Then the card was returned to the clerk to have the rates for the work entered onto a special cost sheet. It then was filed in numerical order for future reference [6]. However, there were exceptions to every rule—Charles Lukens Huston immediately noted, in his instructions, that if there was immediate need for machine shop work to keep the plant running, this routine could be ignored.

DRAWINGS AND BLUEPRINTS

In nineteenth- and early twentieth-century engineering, drawing was an important part of technical communication. It was regularly taught in engineering programs and was a required course in the sophomore and junior years at West Point in 1832 [7, p. 30]. Combined with descriptive geometry, drawing made up 32% of the required curriculum for the mechanical engineering degree at the Massachusetts Institute of Technology in 1867 [8, p. 22]. The *Transactions of the American Society of Mechanical Engineers* published at least five papers about drawing between its founding in 1880 and 1901. Drawing was taught because engineers used it to clarify their ideas and communicate with others. Drawings were also used for construction planning; often the drawings themselves became the specifications for the work that followed. As Eugene Ferguson notes: "When the designers think they understand the problem, they make tentative layouts and drawings, analyze their tentative designs for adequacy of performance, strength, and safety, and then complete a set of drawings and specifications. The second

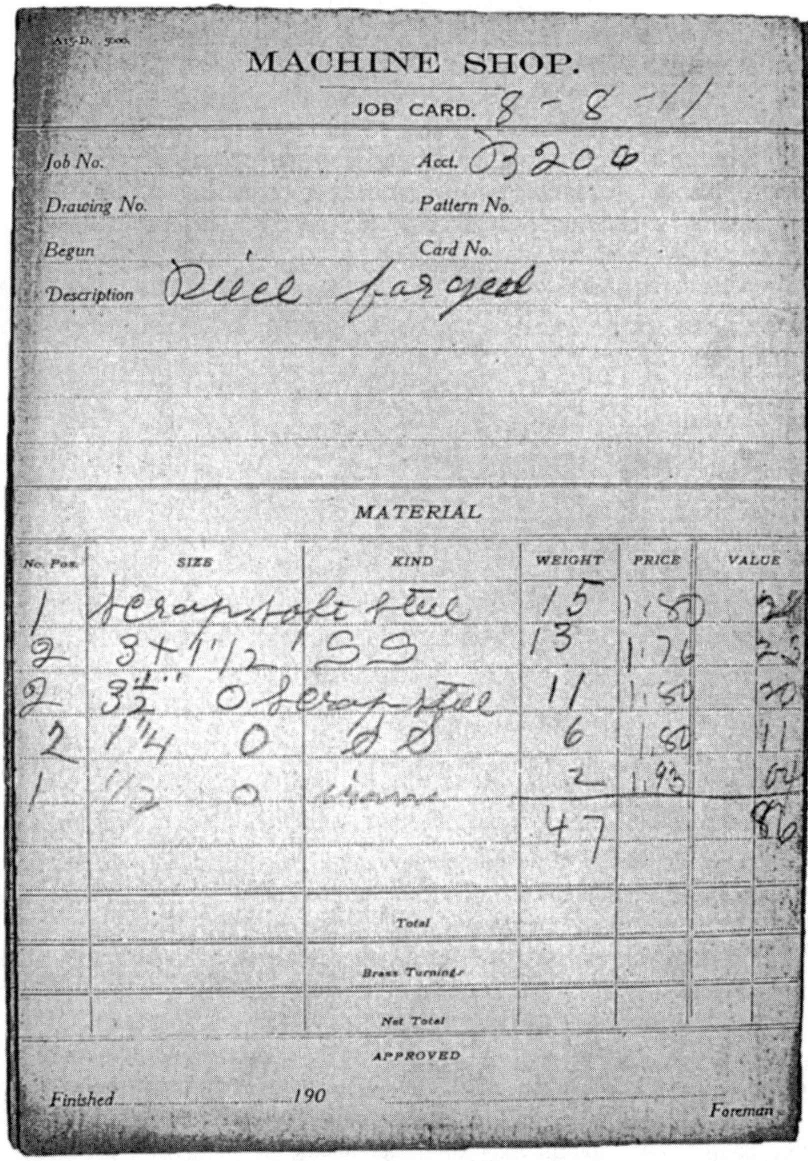

A15-D. 5000

MACHINE SHOP.

JOB CARD. 8-8-11

Job No. *Acct.* B204

Drawing No. *Pattern No.*

Begun *Card No.*

Description Reel forged

MATERIAL

No. Pcs.	SIZE	KIND	WEIGHT	PRICE	VALUE
1	Scrap soft steel		15	1.80	27
2	3 × 1 1/2	S S	13	1.76	23
2	3 1/2"	O Scrap steel	11	1.80	20
2	1 1/4	O S S	6	1.80	11
1	1/2	O [illegible]	2	1.93	04
			47		86
		Total			
		Brass Turnings			
		Net Total			

APPROVED

Finished .... 190 .... *Foreman*

Figure 6. **Machine Shop Job Card (1911).** This was one of the systemized ways of keeping track of various processes that were occurring throughout the works.

process revolved around the finished drawings and specifications. Those drawings and specifications will be the formal instructions that guide their work" [9, pp. 2-3]. In some companies, engineers made drawings that would then be mounted on cardboard and clamped in place in front of the machinists as they worked. Multiple drawings were necessary, not just for distribution, but for permanent documentation, because if anything went wrong they could go back to those drawings in order to see how to rebuild. Thus, the storage and cataloguing of the drawings was an important issue as well.

Most nineteenth century articles about the drawing process recommend that only one person be in charge of the indexing and filing of the drawings, sort of a librarian of drawings. Indexing was important because it contained the history of the machinery and architecture. Many industries carefully stored their indexed drawings in flat drawers about two inches deep, sometimes in a fireproof vault [10]. The wooden patterns for cast iron parts were also made using drawings and were numbered, indexed, and stored in a separate building. Many engineering organizations advocated drawing offices as a fulcrum for handing out work since "dimensioned plans had become an essential tool in the professional practice of mechanical engineers the world over" [11, p. 196].

At Lukens Steel, they did not use drawings as a method for distributing work or as an organizational system—they used drawings continually, but the organization of their plant was somewhat haphazard. By and large, each unit of the plant operated independently, and the necessary communications were passed back and forth by an informal network of managers, foremen, and workers. Lukens used drawing in a variety of different ways for different purposes. Some were what Ferguson called "talking" drawings, some were "thinking" drawings and some were "prescriptive" drawings. The talking and thinking drawings were everywhere—in the correspondence, on scraps of paper and in bound books. Their prescriptive drawings were on starched linen or they were blueprints with multiple copies, which grew in size and complexity over the years.

As John Brown stated, "novel designs generally originated with an oral description or a written specification from a customer or with a novel idea from the proprietor or his engineering talent. A lead draftsman-designer then translated this original concept into a rough sketch. Fundamental to the innovative process, such sketches often passed through many iterations as the designer, the firm's proprietor and/or engineer, and the customer came to agree on the ideal combination of design elements, parts, and proportions" [11, p. 210]. This is basically how it worked at Lukens Steel. Many people had ideas that were communicated via preliminary sketches, many of which were probably not saved. If an in-house drawing had to be done formally, it was sent to Alfred Goodfellow, a mechanical engineer who had that skill. Some of the blueprints in the Lukens archives are from other companies, ordering shapes or trading knowledge, and some of them were made in the plant drawing room.

Drawing was an essential part of the vocabulary of technical communication at Lukens Steel. "Moreover, it is often a collective activity, a communicative practice of negotiation and exchange among participants with different knowledge backgrounds and positions" [12, p. 22]. At Lukens Steel, the engineers, managers, and workers all used drawings to perform their daily work. There were drawings for order specifications, drawings for plant layout and design, drawings for flanged and forged products, drawings as templates for wooden patterns, and drawings as specifications for complex machinery. The foremen used drawings to communicate problems (shown earlier in Figure 5 and later in Figures 10 through 13), and Charles Lukens Huston often used drawings to communicate his ideas while he planned the configuration of new mills. Drawings of all shapes and sizes were scattered throughout the plant, and Lukens saved many of them. The earlier drawings were done on "drafting linen," sized with starch, creating a translucent surface with the texture of oilcloth. However, since multiple images could be made with the blueprinting process, it began to predominate (Figure 7). "Blueprints are produced by placing a transparent (or sufficiently translucent) sheet containing an opaque image in contact with a piece of paper sensitized with potassium ferrocyanide and ferric citrate, and exposing the image-bearing sheet to light. Those areas of the sensitized sheet protected by the lines of the image remain white, while the exposed parts turn blue. After exposure, the blueprint paper is washed to remove residual chemicals" [13, p. 155]. As a copying method, blueprinting predated carbon paper, and sometimes it was used for graphs and text as well.

Lukens Steel used an index for their drawings up until approximately 1915. When they began the index, it was a considerable undertaking; many drawings already existed and had to be sorted into some kind of order. The organization of the index was never entirely satisfactory, and it was never complete. It was separated into alphabetical sections, each letter representing one part of the plant, one system in the plant, or one type of structure in the plant. These categories include "gas producers," which were then subdivided into old plant, universal mill, slab mill, not in use, foreign, open hearth #2, and 140 mill. Similar sections were filed under different letters for boilers and heaters, the 112-inch mill, the 84-inch mill, the flanging department, steel plants, buildings, and shops. There are also sections listing the drawings used for air compressors, trestles, coal and scale conveyors, and "frogs-crossings and track fittings." Most of these sections are further divided into subsections. The index was a large and well-used book, covered with black grime, fingerprints, and handprints. It was kept in a functioning part of the plant rather than an office, as can be seen in the dirt on the cover and most of the pages (Figure 8).

The drawings were numbered roughly sequentially, with space left between each drawing for further revisions. If there was no number available for a new drawing, a letter was sometimes appended to the number. Many of the pages were largely scripted by a single hand, which means a single person was

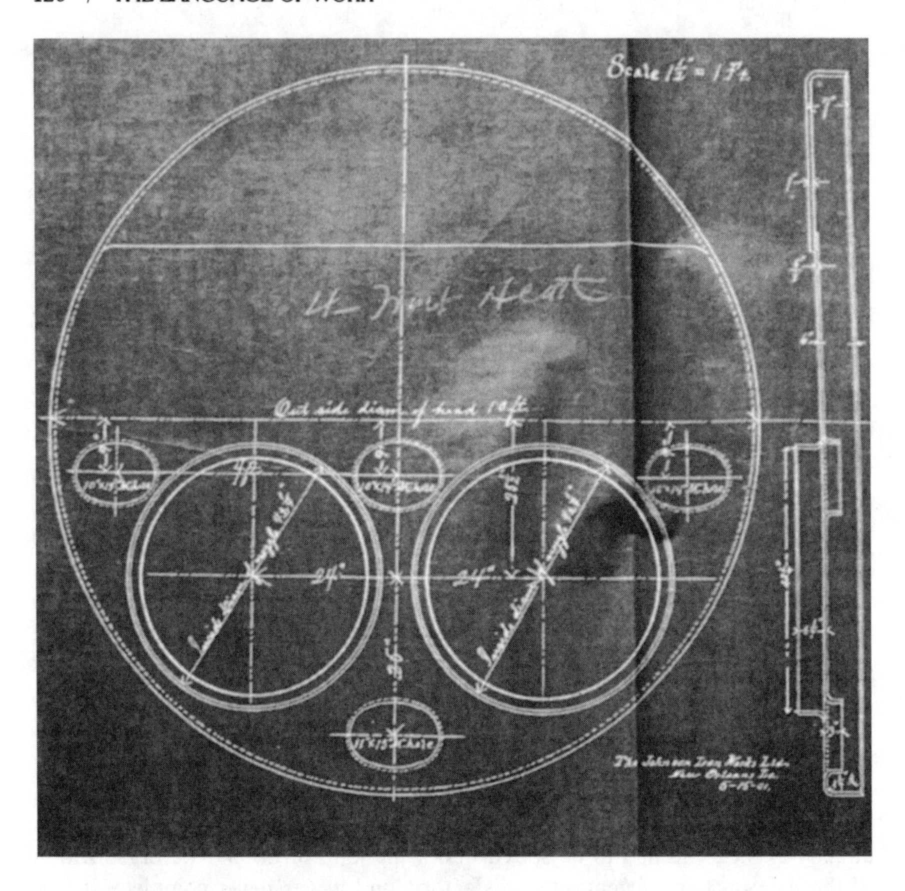

Figure 7. **Blueprint for Boiler Head, The Johnson Ironworks (1901).** This approximately three-foot by three-foot blueprint is one of many from customers specifying the product that they required. Blueprints were used for copying multiple images for engineering, building plans, and writing.

originally put in charge of it. However, additions and deletions were made by others, mixing the ornate clerk's script with printed letters, cross-outs, scrawls, and sometimes shorthand. Drawing titles were added, at different times by different hands, sometimes between the lines, changing the system numbering from sequential numbers to numbers and letters (e.g., 1, 2, 2a, 3). Many notes about the drawings are in the right-hand column, giving additional information, while many titles are crossed out, saying, "Not in Use." Sometimes notes are in shorthand, since shorthand is used throughout the Lukens archives and can be seen in letters and notes as well (Figure 9).

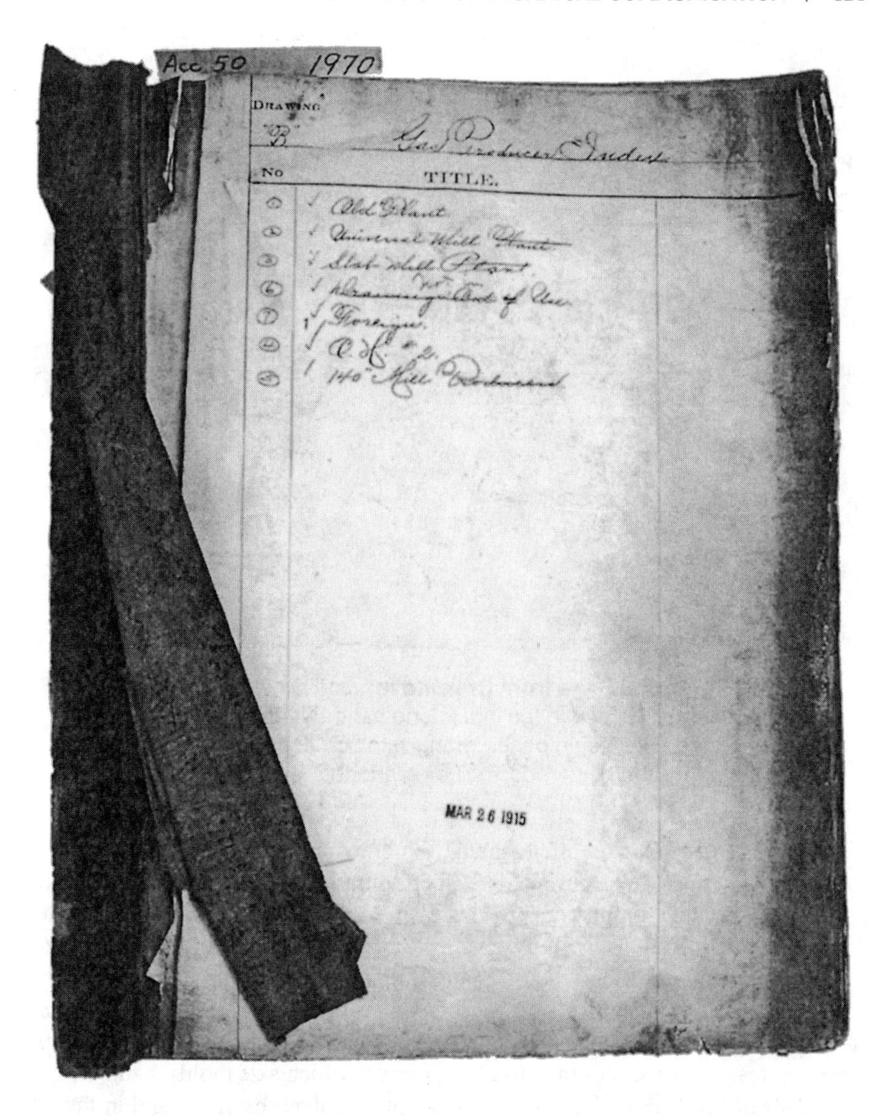

Figure 8. **Drawing Index (1915).** This book, begun by a single scribe and added to by many others over the years, contained a list of all the drawings at Lukens Steel. It was used so frequently that it is missing its cover and many of the pages are marked with black dust, fingerprints, and handprints.

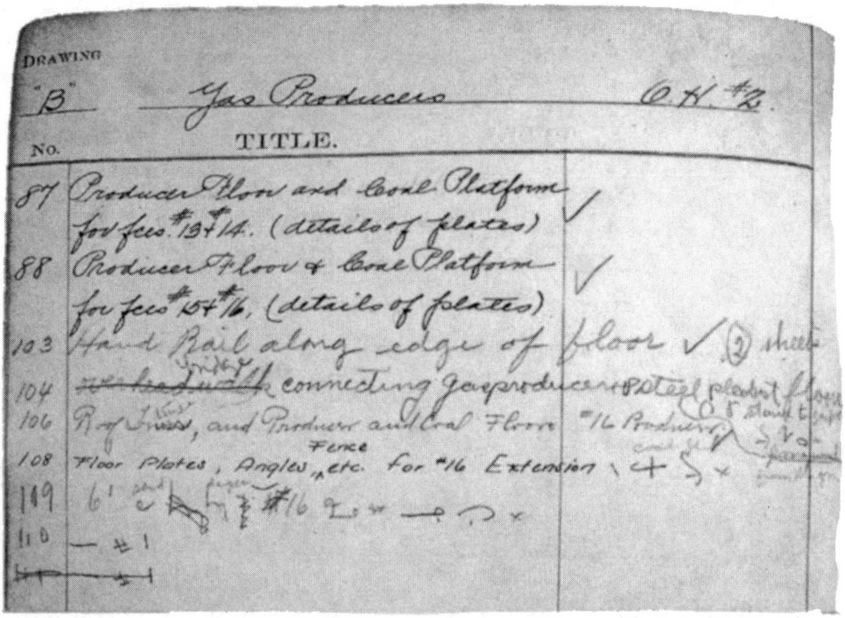

Figure 9. **Partial Page from Drawing Index.** This excerpt shows the variety of scripts used in the index: one hand began the project in ink and added to it later in pencil, along with additions and correlations in other hands, including shorthand notes.

Charles Lukens Huston often "thought in drawing" and many sketches are on his letters and notes as he designed new configurations of machinery for the mills (Figure 10). Other employees, by necessity, also included drawings in their written communications, such as when E. Barnes devised a set of different-sized ingot molds and described their varying outputs: he used slips of paper with a drawing to show the dimensions and quantitative information about the plate rolled from it (Figure 11). These multimedia communications were part of the ongoing testing process as they tried to discover which size molds made the best rolled steel. Literacy in drawing and in writing evolved hand in hand in the late nineteenth and early twentieth centuries. Small drawings within documents were often necessary to explain different shapes and properties (Figure 12). Sketches of equipment and processes were a bridge between the physical and the abstract; they were one step away from a worker standing in the plant, pointing to the parts of a machine, and explaining what it was doing. Drawings had nearly the same effect as intimate, tacit knowledge exchange, but they could be sent across distances, used at different times, and shown to multiple people. Drawings were an integral method of communication in the early twentieth century as the managers and foremen explained their ideas (Figure 13).

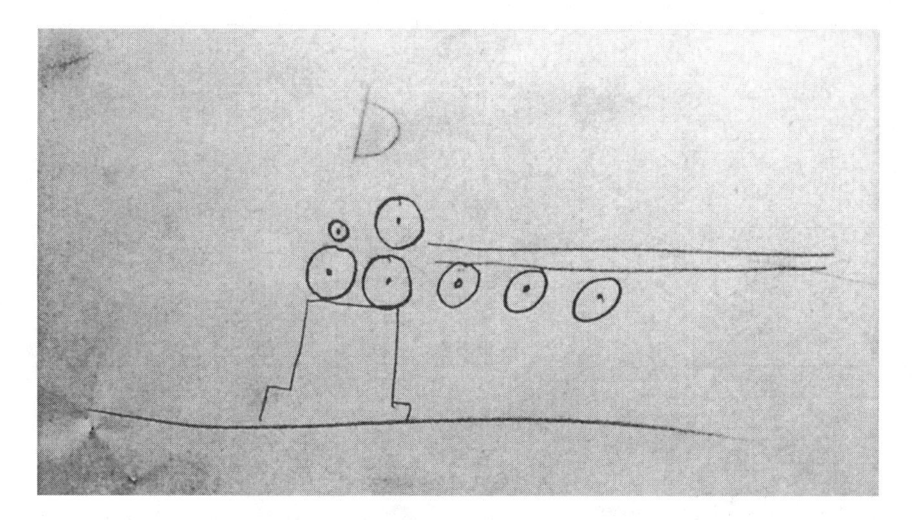

Figure 10. **Sketch in Charles Lukens Huston's Papers.**
This undated drawing was pinned to another entitled
"Emmett Moore's Plate Dumping Scheme."

As Ferguson noted, engineers such as Charles Lukens Huston often thought by drawing. Huston often went to the various parts of the mills and watched the men as they worked. When he wasn't complaining about their lackadaisical behavior, he was taking in the entire industrial process—how each machine worked, how they worked together, and what processes connected the output from the machines—and he looked for areas that could be improved. His greatest strength in finding areas for improvement was in theorizing improved rolling mills. He watched, saw how they worked, drew sketches of how they could work, discussed them with people, and implemented many of these improvements into the mills. He also patented several of the improvements, and they were adopted by other rolling mills. This process of thought, communication, and innovation used and resulted in drawings.

For instance, the Lukens plant made specialized shapes called "heads," which were end caps for boilers and other large circular objects. At the end of the nineteenth century, most heads were still made by hammering iron or steel over a template. Dr. Huston and his sons saw a new "hot spinning" machine used to make heads and ordered one in 1885. Huston then designed several improvements to the new machine, including guides for outside rollers, which became standard in the industry [14]. There exists an undated drawing of the spinning machine called a "Roller Dishing Device," unlisted in the index (although it is listed under a different number), which was an early iteration of what later became commonly known as a flanging machine (Figure 14). Later, Lukens had several flanging

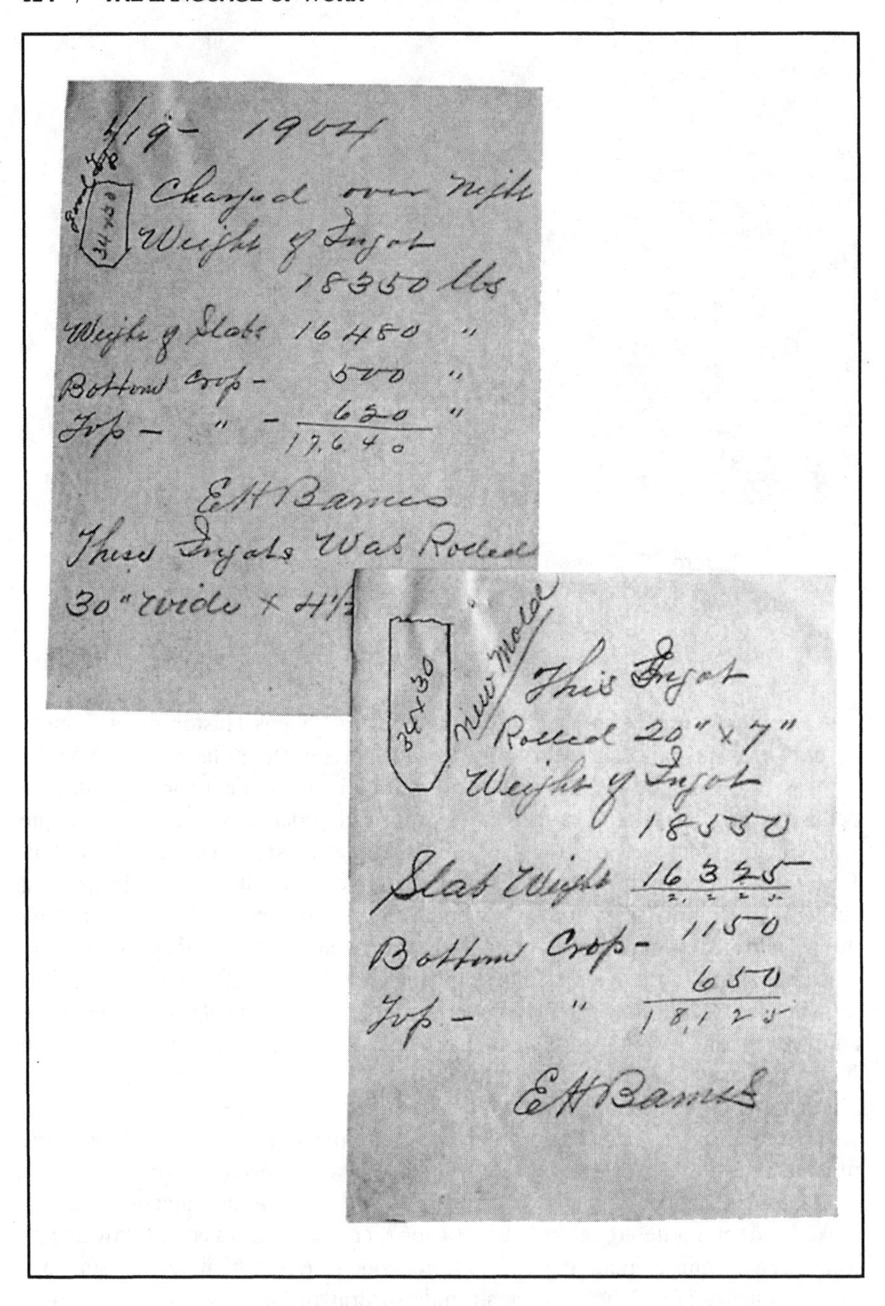

4/19 – 1904
Changed over Night
Weight of Ingot
18350 lbs
Weight of Slab 16480 "
Bottom Crop – 500 "
Top – " – 620 "
17,640
E H Barnes
These Ingots Was Rolled
30" wide × 4½

New Mold
This Ingot
Rolled 20" × 7"
Weight of Ingot
18550
Slab Weight 16325
Bottom Crop – 1150
Top – " 650
18,125
E H Barnes

Figure 11. **Visual, Verbal, and Quantitative Communication.** These notes are an example of how the workers used writing, drawing, and calculations simultaneously.

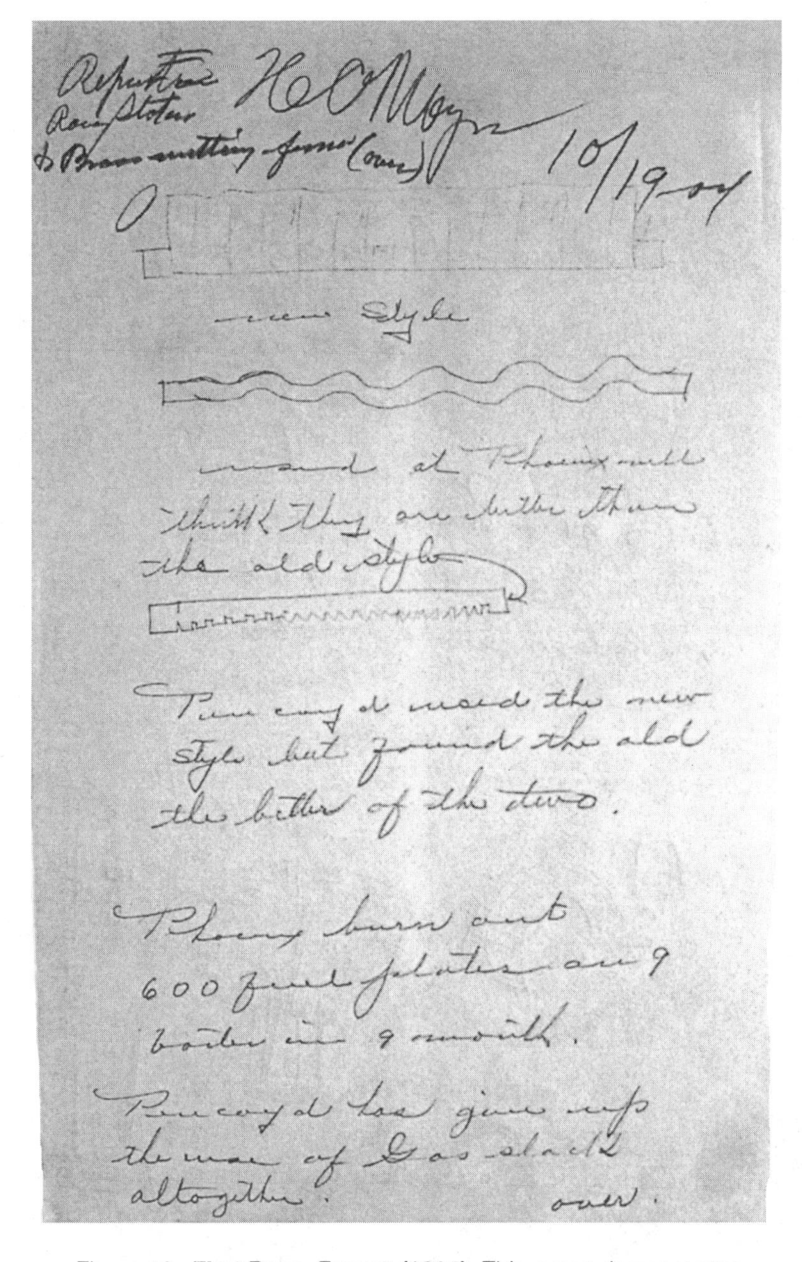

Report re Roll Stoker & Brass [illegible] forms (over) H C Mgr 10/19-04

new Style

used at Phoenix mill think they are better than the old style

Pencoyd used the new style but found the old the better of the two.

Phoenix burn out 600 fuel plates on 9 boiler in 9 month.

Pencoyd has given up the use of Gas slack altogether.

over.

Figure 12. **First Page, Report (1904).** This report demonstrates how the author moved from expressing technological ideas visually to expressing them verbally and then back again. The title and date is in Charles Lukens Huston's hand.

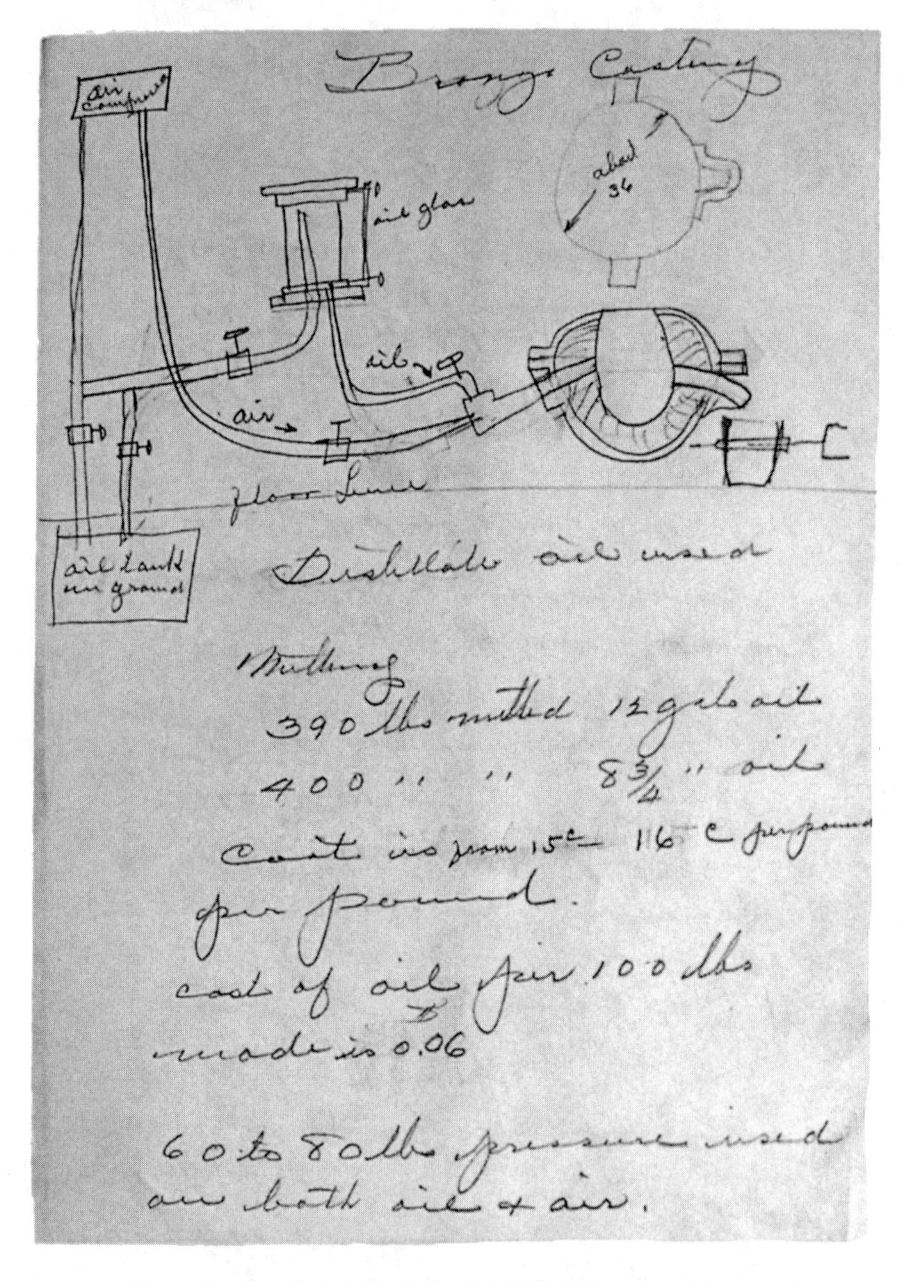

Bronze Casting

Distillate oil used

Melting
390 lbs melted 12 gals oil
400 " " 8 3/4 " oil
Cost is from 15c — 116 c per pound
per pound.
cost of oil per 100 lbs
made is 0.06

60 to 80 lbs pressure used
on both oil & air.

Figure 13. **Second Page, Report (1904).** This page of the report shows a visualization of process and includes quantitative calculations containing information about the problem they were trying to solve.

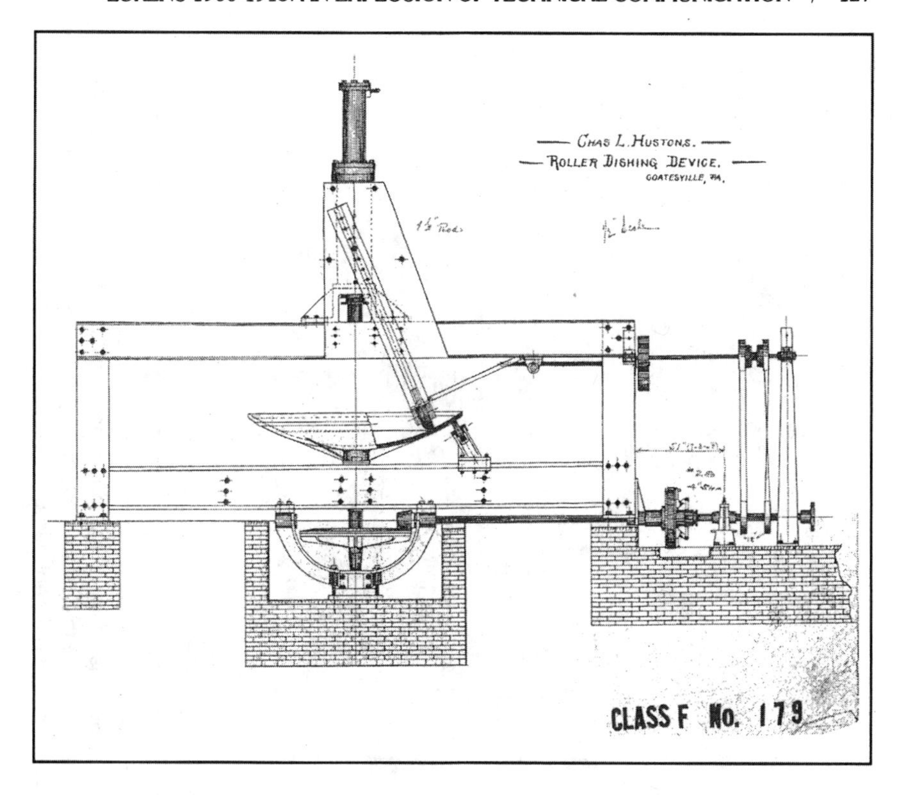

Figure 14. **Roller Dishing Device by Charles Lukens Huston.** This undated drawing was done on sized linen and is approximately three-feet by four-feet. It was probably drawn in connection with Huston's work on adding guides to an earlier device.

machines that were so complex that no single drawing would suffice to describe its specifications; blueprints with multiple drawings had to be made of the interacting parts (Figure 15). Drawings are more closely connected to the physical reality of an object and any writing about it and thus are especially useful in engineering. As Ferguson wrote, "The affinity of engineering for art has been masked by the rise of the physical sciences, but the successful practice of engineering will always be shaped by the disciplines of art" [9, p. 72].

THE EMERGENCE OF TESTING

Although Lukens Steel was active in testing from 1875, when Dr. Charles Huston purchased a testing machine and ran two scientific tests that he published in the *Journal of the Franklin Institute*, little evidence of this early testing

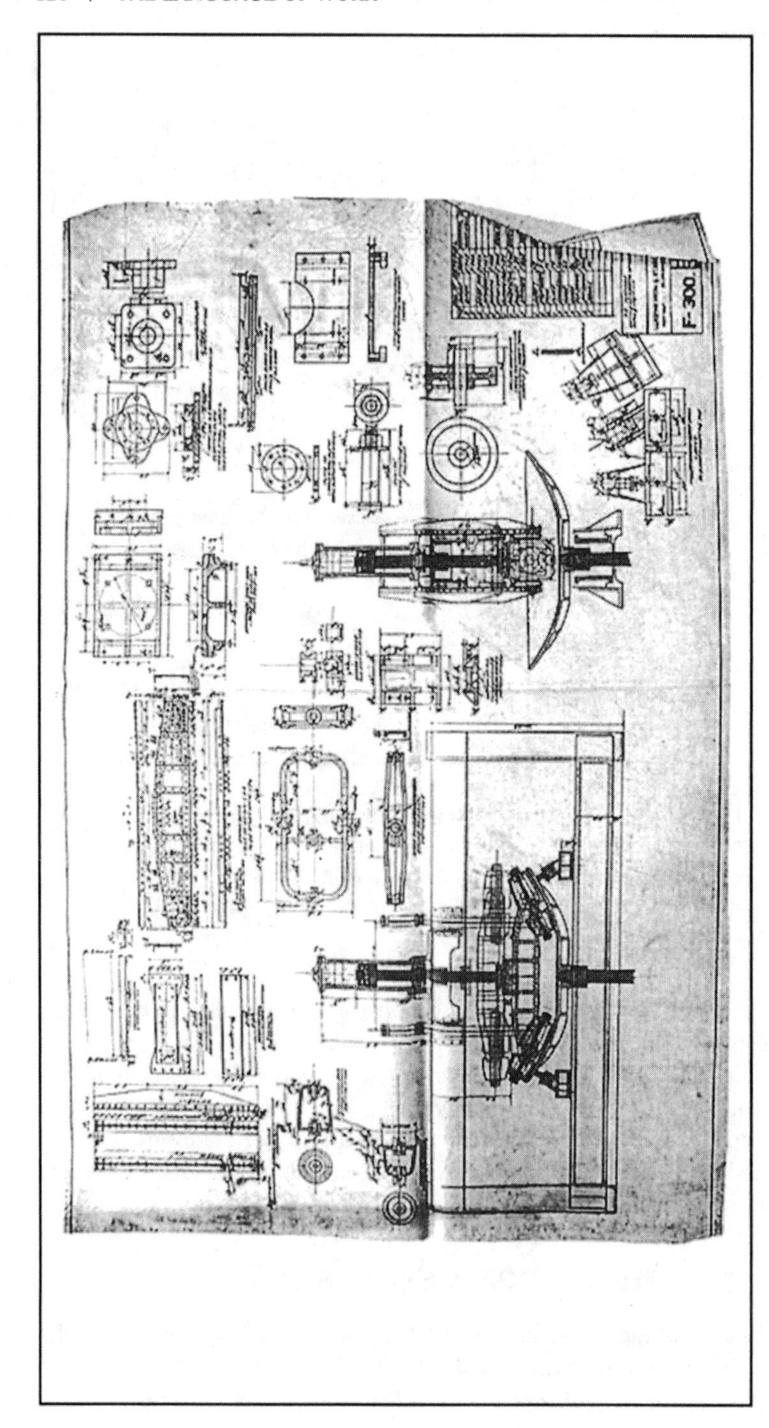

Figure 15. **"#2 Flanger, Details of Dishing Device" (1915).** This drawing includes more detailed information about the increasingly complex machinery: eventually it was not possible to describe machinery without multiple drawings from different views and of different parts.

remains: until the turn of the century, the business owners didn't find this material worth saving. Even complex scientific experimentation was carried on with minimal written technical communication, and the necessary steps occurred prediscursively. From the 1870s the U.S. government required that the tensile strength be stamped on plate metal that was going to be used as boiler plate in railroad and steamboat engines (Dr. Huston participated in the setting of these standards). Thus, from early on, testing was part of the manufacturing process at Lukens Steel. However, it wasn't fully documented until 1900.

At the turn of the century there was an explosion of documentation about testing (as well as in the other parts and processes in the plant). New social discourse communities sprang up as workers took on new tasks to track and understand visible and unseen industrial processes and their products. As these workers attempted to control manufacturing output, names emerged for them such as Engineer of Testing, Metallurgist, Chief Chemist, Inspector Foreman, and Fuel Economist. One thing they all shared was that they used writing to collect, store, and communicate their results.

There were four general types of testing at Lukens Steel after 1900: in-house testing of plant equipment (generally of boilers and stacks); testing as part of the manufacturing process (prior to inspection); testing as scientific experimentation (often resulting in presentations and publications); and testing as part of a social discourse community that emerged in order to create industry standards (composed of government, industry, and academics). After 1915 testing continued to gain importance in all four areas, and the amount of documentation about it increased exponentially.

In-House Testing of Plant Equipment

Lukens periodically tested its own machinery and systems. The testing documentation began in 1899 when they collected copies of boiler tests from other plants, such as Pencoyd Iron Works and General Electric, to use as templates for their own tests. Frequently Lukens collected knowledge from other plants, and other plants came to Lukens to observe their procedures too. The general attitude of sharing knowledge freely can be seen throughout the years covered in this book. Lukens then used the templates they received from other companies and had an outside consultant set up their testing procedures. One part of this process was that they began to test their own engines with indicator cards attached to a drum or another mechanical arm in order to record the movements of pistons in a steam engine and thereby calculate the power output (Figure 16). Indicator cards were a mechanical means of recording pressure on paper: the card was wrapped around a cylinder that had a stylus attached to a piston type pressure gauge. The area inside the indicator diagram is used to calculate the power output (Figure 17). Lukens had tests of this sort done on their steam engines in the 84-inch mill, the 120-inch mill, and the universal mill in 1900 and 1903. They

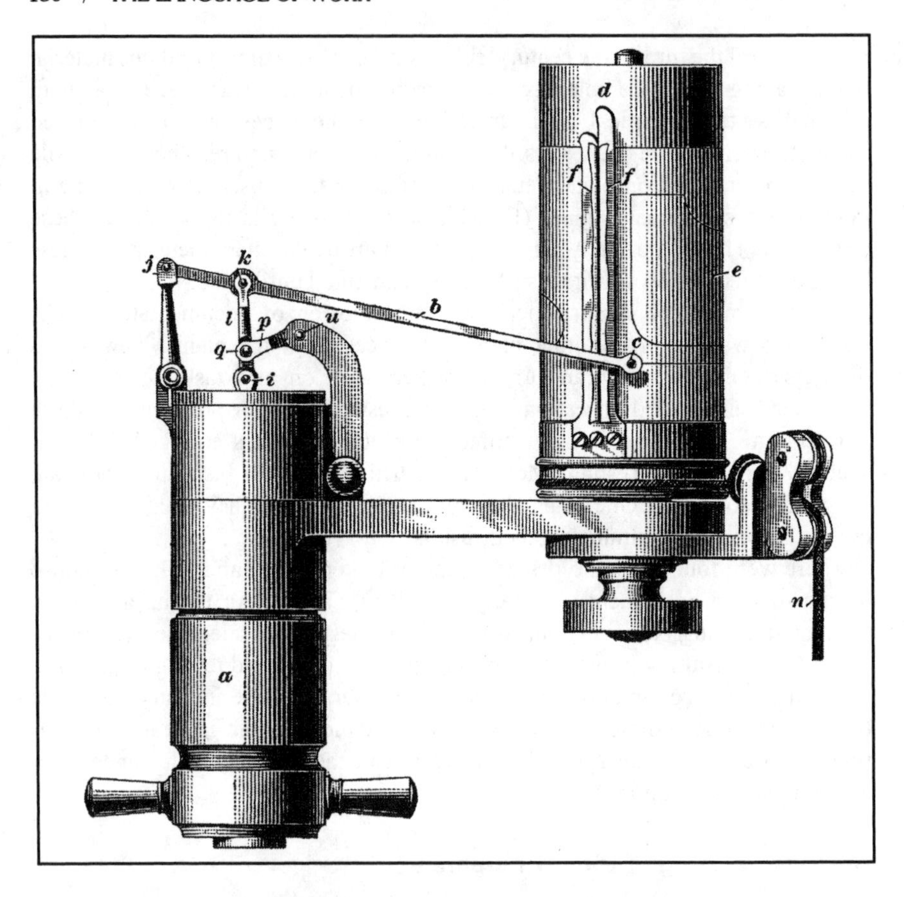

Figure 16. **Indicator Card Device for Testing Engine Power.** This device was attached to an engine to measure its power output. A card was attached to the drum and a pencil, connected to a pressure-measuring device, drew a diagram on the card [15].

also frequently tested the temperature of the air leaving the stacks, the quality of the water, and any other aspect of the plant they needed information about. All of this testing resulted in documentation.

Soon Lukens developed its own forms for testing boilers. A multipage official report titled "Testing Department: Log of Boiler Test," with several subsections: "Report of Boiler Test," "Log of Boiler Test," and "Coal and Water Log." These tests required observers to take temperatures and other readings at set intervals so that they could be combined in the report. The printed lab reports required such information as the type of boiler, the manufacturer, name of tester, kind and size of coal, moisture in coal, quality of steam, total water evaporated, steam

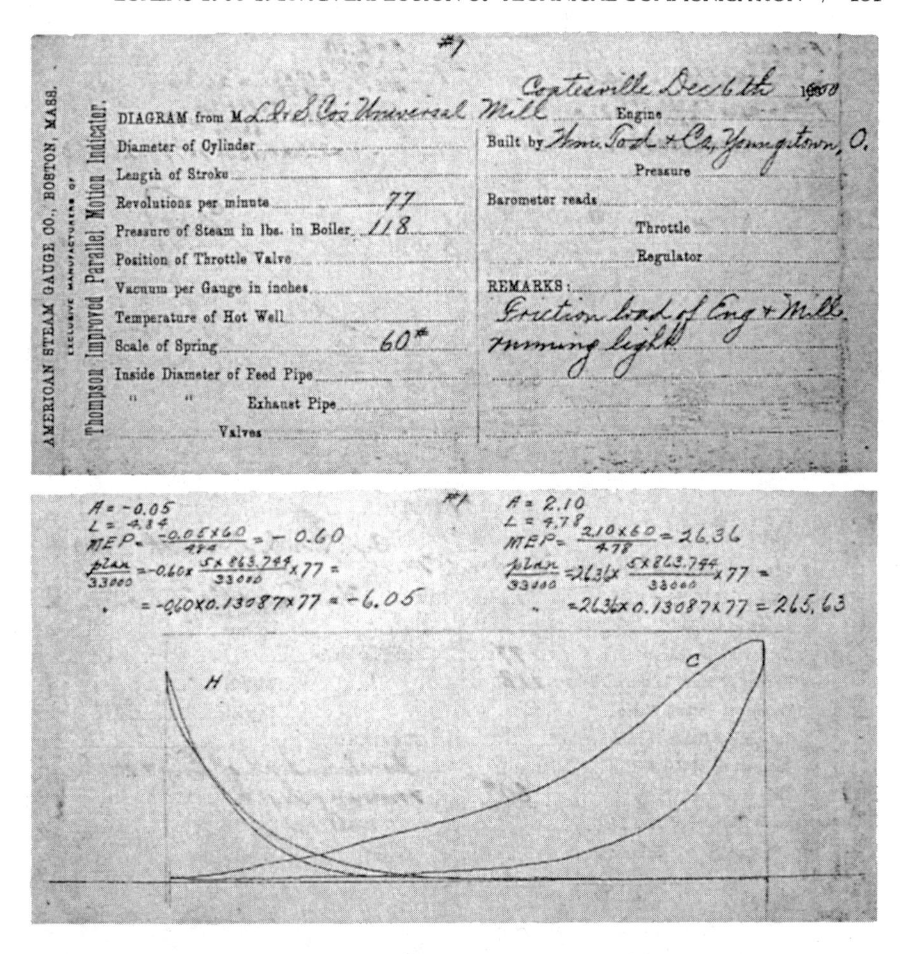

#1

AMERICAN STEAM GAUGE CO., BOSTON, MASS.
EXCLUSIVE MANUFACTURERS OF
Thompson Improved Parallel Motion Indicator.

Coatesville Dec 6th 1900

DIAGRAM from M L. I. & S. Co's Universal Mill Engine

Diameter of Cylinder .... Built by Wm. Tod & Co, Youngstown, O.

Length of Stroke .... Pressure ....

Revolutions per minute 77 Barometer reads ....

Pressure of Steam in lbs. in Boiler 118 Throttle ....

Position of Throttle Valve .... Regulator ....

Vacuum per Gauge in inches .... REMARKS:

Temperature of Hot Well .... Friction load of Eng + Mill

Scale of Spring 60# running light

Inside Diameter of Feed Pipe ....

" " Exhaust Pipe ....

Valves ....

#1

A = -0.05
L = 4.84
MEP = -0.05×60/4.84 = -0.60
plan/33000 = -0.60 × 5×863.744/33000 × 77 =
= -0.60 × 0.13087 × 77 = -6.05

A = 2.10
L = 4.78
MEP = 2.10×60/4.78 = 26.36
plan/33000 = 26.36 × 5×863.744/33000 × 77 =
= 26.36 × 0.13087 × 77 = 265.63

Figure 17. **Front and Back of an Indicator Card (1900).** The rotation of the drum resulted in a diagram that could then be used to calculate the power output. This method of testing could show inefficiencies that could then be corrected.

per pound of fuel, etc. (Figure 18). Although the summary data from the consultant was expressed numerically, when the employees at Lukens took over the reports, the summary was a narrative.

> The actual horse power developed, by this boiler, as [crmp] and with the builders rating served so high, that it was deemed best to run a check test. This test, while it was impossible to finish by reason of the clogging of the grate-door to bad coal which made it impossible to clear the first and necessitated shutting down of the boiler, proved the figures in the test as were reported were correct.

75

5974

TESTING DEPARTMENT—LUKENS IRON AND STEEL CO., COATESVILLE, PA.

REPORT OF BOILER TEST.

Made by P. C. Haldeman

Date July 30, 1903

Kind of Boiler Horizontal Return Tubular *Manufactured by* Coatesville Boiler Wks.

	Duration of trial ... Hours.	8
DIMENSIONS.	Grate surface: length ... ft., width 5 ft. Sq. ft.	37
	Water heating surface ... "	1866
	Superheating surface ... "	
	Height, chimney Induced Draft Ft.	
	Ratio of draft area to grate 11 to 37 Sq. ft.	.3
	Area, chimney ... "	
	Tubes: Diameter 3" No. 118	
	Ratio heating to grate surface	50.4
PRESSURE.	Barometer 14.49 Inches mercury.	2953
	Steam gauge ... Pounds.	118.06
	Absolute steam pressure ... Pounds.	132.5
	Draught gauge, near damper ... Inches water.	.8
	" " in ash pit ... " "	
TEMPERATURE.	External air ... Degrees F.	87.6
	Boiler room ... "	119.
	Escaping gasses ... "	449.5
	Furnace ... "	
	Feed-water ... "	170.7
	Steam ... "	
FUEL.	Kind of coal Bituminous	
	Locality "Winton" Clearfield (?)	
	Size of coal Run of Mine	
	Fixed carbon ... Per cent.	60.45
	Volatile matter ... "	25.76
	Moisture ... "	1.46
	Ash ... "	12.33
	B. T. U. per pound coal	13598
	" " combustible	15660
	Carbon in ash ... Per cent.	20.99
	Total coal consumed ... Pounds.	8906
	Total Wood ... "	
	Moisture in coal ... Per cent.	1.46
	Dry coal consumed ... Pounds.	8786
	Total refuse, dry ... "	1221
	Total refuse, dry ... Per cent.	13.9
	Total combustible ... Pounds.	7574
COMBUSTION PER HOUR.	Dry coal ... "	1098.25
	Combustible ... "	947
	Dry coal per square foot of grate ... "	29.7
	Combustible per " " " ... "	25.6
	Dry coal per " " Heating surface "	.59
	Combustible per " " " " "	.50

FLUE GAS.	CO_2 ... Per cent of volume.	5.8
	CO ... " " "	.1
	N ... " " "	80.4
	Free oxygen ... " " "	13.7
	Air ... Excess coefficient.	
	Quality of steam ... Per cent.	98.58
	Superheat ... Degrees.	
TOT. WATER.	Total weight water used ... Pounds.	82785
	Total evaporated, dry steam ... "	81609
	Factor of evaporation ...	1.086
	Equivalent evaporated from and at 212° ... "	88627
WATER PER HR.	Amount used ... Pounds.	10348
	Evaporated dry steam ... "	10201
	Equivalent evaporated from and at 212° ... "	11078
ECONOMIC EVAPORATION.	*Per Pound of Fuel.*	
	Actual, per lb. dry coal ... Pounds.	9.3
	Equivalent from and at 212° [dry coal] ... "	10.09
	Per Pound of Combustible.	
	Actual ... Pounds.	10.77
	Equivalent from and at 212° ... "	11.70
RATE OF EVAPORATION.	*Per Sq. Foot of Grate per Hour.*	
	Actual, from feed-water temperature ... Pounds.	275.7
	Equivalent from and at 212° ... "	294.4
	Per Sq. Foot of Water-heating Surface per Hr.	
	Actual ... Pounds.	5.46
	Equivalent from and at 212° ... "	5.94
	Per Sq. Foot of Least Draught Area per Hr.	
	Actual ... Pounds.	940.7
	Equivalent from and at 212° ... "	1007.1
H. P.	*On basis 34½ pounds equiv. evap. per hour. H. P.	321
	Builders' rating ... "	150
	Ratio of commercial to builders' rating ...	2.14
Heat Balance		B.T.U. / Per ct.
	Heat generated per pound of coal ... 13598	100
	Heat absorbed " " " ... 9743	71.6
	Heat lost in flue " " " ... 3298	24.2
	" " incomplete combustion lb. of coal 12	.8
	" " by radiation pound of coal ... 107	
	" " in ashes " " ... 439	3.4
	(a) Efficiency of Boiler ...	72.1
	(b) " " boiler and grate ...	71.6

NOTE.—Actual evaporation signifies the evaporation from feed-water temperature to dry steam at gauge pressure. It is apparent evaporation corrected for calorimeter determination.
* Standard commercial H. P. (a) Computed from combustible. (b) Computed from coal.

Figure 18. **Page from Lukens Boiler Test Template (1903).** Lukens had many forms printed like this which could serve repeated functions. Although only the boiler tests from 1900 to 1904 are preserved, they document the procedure.

The feed water was much lower than normal on account of one of the economizers being out of service.

Both the quality of the steam and the draft seemed to vary some what,—and at—times when the engines were draining hard—a good deal of water was carried over through mains.

The flue-gas temperature was low except = at the end of the test when the fires were forced harder.

This test was made as much as possible under actual running conditions [16].

Specialized products required specialized working conditions, which changed continually and could not be reduced, as yet, to a predictable process.

Stack tests were also done on the later pages of the boiler tests. They, too, had the date, the furnace, and the name of the tester across the top. There were columns for time, draft gauge, and flue temperature. The tests were taken every 15 minutes from 9 A.M. to 10:30 P.M. and the results were summarized at the end; for example, "Averaging a discharge of 2653 Cu. Ft. per second at average temperature of 1156 degrees F." These tests were performed on different flues on different dates in the fall of 1904. They also tested the temperature of the pits from time to time, taking readings every five minutes and jotting them down on six-by-nine paper. They also took averages of the gas producers in down-takes, the temperature of the air in the airbox, and gas in the main lines in separate documents [6]. In-house testing was an integral part of the industrial culture at Lukens Steel, and the documentation that they developed allowed analysis of problem areas and provided clues to solve them.

Testing as Part of the Manufacturing Process

The testing department was a necessary part of the manufacturing process because almost every type of steel that Lukens made was for use in railroad or marine boilers and commissioned either by the railroads or government, both of whom had strict standards with which they had to comply. Every plate was stamped with the heat number, a letter for the ingot from which it was rolled, and the tensile strength. There was a "checker" at the last stage in the process, who was responsible for checking off each order by size, which was a vital piece of data for the shipping department. This checker also caught all sorts of mistakes, like being sheared wrong or recorded wrong, together with other inconsistencies [6]. Thus, each mill had a recording office. The Testing Department was responsible for conducting and coordinating inspections before shipping and also for interacting with customers regarding rejected product.

In the late nineteenth and early twentieth century, the testing department was run by long-time employee Howard Taggert. There was a complex system of numbering the path of each piece of steel produced. It started with the numbering of each heat in the open-hearth plant and then assigning a letter, after the heat number, to each ingot poured from the heat. After rolling, each plate had a portion sheared off, which was tested for tensile strength. Each plate and test piece was stamped with the heat number, the ingot letter, and the tensile strength. When W. G. Humpton later became the testing engineer, he complained, "Think good deal of trouble could be eliminated if cut out some stamping. There is so much stamping in test piece that it really invites error at the mill" [17, 12/23/18]. Marine boiler plate manufactured for the government had to undergo a bend test as well. Although the testing process was complex, it served Lukens well for years since they were able to identify each plate that they sold and, therefore, find the cause of problems. Sometimes a customer complaint was due to customer error rather than plate defects and, if it was a repeated defect, they could attempt to find the source and change their procedures.

Charles Lukens Huston was in continual contact with the testing, chemical, and sales departments concerning a variety of questions, most frequently about rejected plates. As seen in the section about interplant communication above, communication regarding these questions often took place by folded notes (see Figures 5, 12, and 13). In one instance, Huston wrote a note to Taggert to ask, "What was the bend test in enclosed, and how did it fail?" Taggert wrote back on the note itself:

> We made cold bending test on half of broken test pce. & it failed—This occurred while I was in Washington—the boys could not persuade Inspector to allow us to prepare a regular bending specimen—when I got home the Inspector had sent in his report & was unwilling to recall it—are preparing specimens for bending on subsequent shipments. HT [6, 2/9/06].

Meeting the inspector's requirements was difficult and unpredictable, and thus a great deal of time and effort went into the process. In a later operations committee meeting, Huston said, "Don't believe we get one plate out of ten to A.S.T.M. [American Society for Testing Materials] specifications" [17, 12/23/18]. The difficulty of producing steel that met the standards provided an impetus to do scientific testing to make the process more controllable. Thus, it may have been the act of inspection itself, more than the properties inspected, that improved the quality of steel.

Two locomotive manufacturers, Baldwin and the American Locomotive Co., had resident inspectors in Coatesville [17, 5/13/18]. Lukens was under constant scrutiny, both by industry and the government, and these inspections produced reports, both from the manufacturing side and from the purchaser's side, as can be seen in this test report from Baldwin Locomotive (Figure 19). Although there were many other records that were kept on the factory floor, the majority of the documents that were used by the plant had to do with testing. Since the heat and

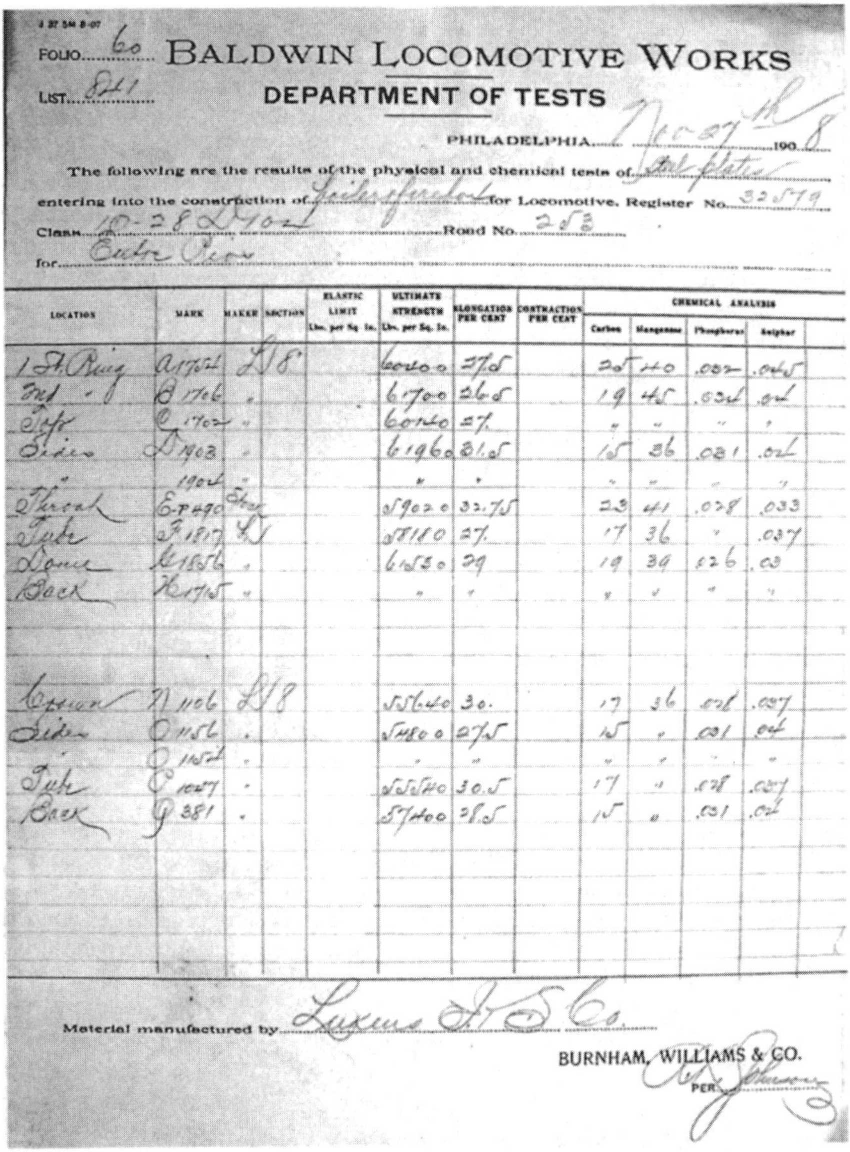

FOLIO 60 **BALDWIN LOCOMOTIVE WORKS**

LIST 841

DEPARTMENT OF TESTS

PHILADELPHIA, Nov 27th 1908

The following are the results of the physical and chemical tests of steel plates

entering into the construction of boiler and firebox for Locomotive, Register No. 32579

Class 10-28 D 102 Road No. 253

for [illegible]

LOCATION	MARK	MAKER	SECTION	ELASTIC LIMIT Lbs. per Sq. In.	ULTIMATE STRENGTH Lbs. per Sq. In.	ELONGATION PER CENT	CONTRACTION PER CENT	CHEMICAL ANALYSIS Carbon	Manganese	Phosphorus	Sulphur
1 St. Ring	A 1704	L.S.			60500	27.5		25	40	.032	.045
Int. "	B 1706	"			61700	26.5		19	45	.036	.04
Top	C 1702	"			60540	27.		"	"	"	"
Sides	D 1903	"			61960	31.5		15	36	.031	.04
"	1904				"	"		"	"	"	"
Throat	E-P490	Shar.			59020	32.75		23	41	.028	.033
Tube	F 1817	L.S.			58180	27.		17	36	"	.037
Dome	G 1856	"			61530	29		19	39	.026	.03
Back	H 1715	"			"	"		"	"	"	"
Crown	N 1106	L.S.			55640	30.		17	36	.028	.027
Sides	O 1156	"			54800	27.5		15	"	.031	.04
"	O 1152	"			"	"		"	"	"	"
Tube	P 1027	"			55540	30.5		17	"	.028	.027
Back	Q 381	"			57400	28.5		15	"	.031	.04

Material manufactured by Lukens I. & S. Co.

BURNHAM, WILLIAMS & CO.

PER [illegible signature]

Figure 19. **Test Report from Baldwin Locomotive Plate Inspectors (1908).** Part of the reason for extensive testing at Lukens Steel was that they were subject to continual inspection from the railroad industry and from government. This is an example of the inspection results from the user's end.

slab numbers were attached to each plate (including test pieces), all of the shapes created by the plate mills could be traced back to the original source if there were defects in the steel. Many of these records are covered with a layer of black dust. The form "Testing Department Report of Tests of Steel" helped to track this information and contained the tensile strength and a chemical analysis as well (Figure 20). Each test record had the name of the furnace; the date; and columns for slab no., melt no.; a chemical analysis of the amounts of carbon, manganese, sulphur and phosphorous; a column for original dimension; the elastic limit; the tensile strength; elongation; reduced dimensions; percent reduction; size of plate; and remarks. This documentation, a natural outgrowth of the record keeping described in the earlier chapter, was a written method of quality control that resulted in data that could be analyzed to solve problems. Lukens' ability to produce and track this amount of detailed information contributed to their ongoing success as manufactures of high-quality specialized steel.

Scientific Testing

Since the open hearth and rolling processes were so complex and many plates failed inspection, Charles Lukens Huston and others were constantly trying to discover the properties in the manufacturing process that would consistently produce high-quality rolled steel. Many of the problems originated with the open hearth process. Originally, Lukens started their own open hearth furnaces so that they could control the chemical content and the shape of the ingots because different-shaped ingots worked best in different mills. Making perfect ingots itself was a complex and unpredictable process:

> A prerequisite to faultlessly finished material is perfect ingots, and by a perfect ingot is meant one free from all cavities or openings and made up of material that is homogeneous throughout. Unfortunately, the natural laws that govern the solidification of the liquid metal operate against both these requirements, and develop the well known natural defects in ingots called piping, blow holds, segregation and crystallization. Added to these are other defects, both incidental and accidental, such as checking, scabs, and slag inclusions [2, p. 459].

It was very difficult to make good ingots, and a great deal of Lukens' time and effort went into testing each melt and using various chemical additives and experimenting with different-sized molds for ingots. Some of these experiments were never published and descriptions of them, with their data (including photographs, microphotographs, and graphs), are saved in the archives. However, Charles Lukens Huston followed in his father's footsteps by publishing some of his results. He continued promoting the idea that tensile strength alone was not a sufficient indicator of the safety of steel products.

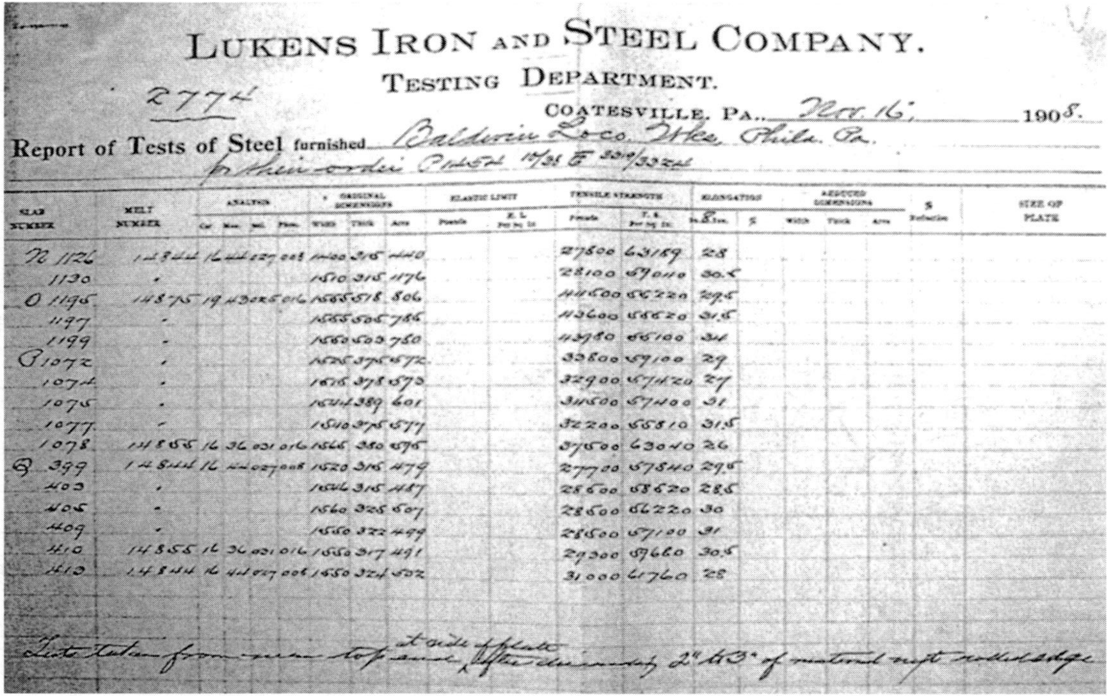

LUKENS IRON AND STEEL COMPANY.

TESTING DEPARTMENT.

2774

COATESVILLE, PA., Nov. 16, 1908.

Report of Tests of Steel furnished Baldwin Loco. Wks. Phila. Pa.
for their order [illegible]

SLAB NUMBER	MELT NUMBER	ANALYSIS (Car. Man. Sul. Phos.)	ORIGINAL DIMENSIONS (Width Thick Area)	ELASTIC LIMIT (Pounds / E. L. Per Sq. In.)	TENSILE STRENGTH (Pounds)	T. S. Per Sq. In.	ELONGATION in 8 In.	%	REDUCED DIMENSIONS (Width Thick Area)	% Reduction	SIZE OF PLATE
N 1126	14844	16 44 027 008	1400 315 440		27800	63189	28				
1130	"		1510 315 476		28100	59040	30.5				
O 1195	14875	19 43 025 016	1555 518 806		44500	55220	29.5				
1197	"		1555 505 786		43600	55520	31.5				
1199	"		1550 503 780		43980	56100	31				
Q 1072	"		1525 375 572		32800	57100	29				
1074	"		1515 378 573		32900	57420	27				
1075	"		1544 389 601		34500	57400	31				
1077	"		1540 375 577		32200	55810	31.5				
1078	14855	16 36 031 016	1565 380 595		37500	63040	26				
S 399	14844	16 44 027 008	1520 315 479		27700	57840	29.5				
403	"		1546 315 487		28600	58620	28.5				
405	"		1560 325 507		28500	56220	30				
409	"		1550 322 499		28500	57100	31				
410	14855	16 36 031 016	1550 317 491		29300	59660	30.5				
413	14844	16 44 027 008	1550 324 502		31000	61760	28				

Tests taken from near top end at side of plate, after [illegible] 2" to 3" of material next rolled edge

Figure 20. **Sample Testing Department Reports of Tests of Steel (1908).** There were columns for slab number, melt number, chemical analysis of the amounts of carbon, manganese, sulphur, and phosphorous; columns for the original dimension, the elastic limit, the tensile strength, elongation, reduced dimensions, percent reduction, size of plate, and remarks.

Huston knew the interior of steel as well as his chemists and the open hearth men. He was especially interested in the segregation that occurred within the ingot– due to uneven cooling, different parts of the ingot had different chemical content and different physical features (Figures 21 and 22). When rolled, therefore, different parts of the plate had different properties as well. He wrote, "it is difficult or really impossible to secure steel that even in one moderate sized plate will have anything like the uniformity generally supposed to exist, because the tensile strength, in almost any one plate of a ton weight or over, will vary 5,000 lbs. or more in tests taken from different parts" [18, p. 182]. Huston set out to demonstrate this in a carefully planned experiment, the results of which were read before the Ninth Annual Meeting of the American Society for Testing Materials in 1906 and then published in two separate journals [18, 19].

Conducting the tests was a collaborative project among workers and managers at the plant. First, Huston asked his chemist, H. G. Martin, to prepare different-sized ingots (9 × 12, 16½ × 18½, 26 × 12, 38 × 8) and have them rolled into plates. Martin then had E. Barnes draw up the specifications (Figure 11). Huston chose ingots of different sizes because the segregation of chemical and mechanical properties occurred differently in each. After pouring and rolling, pieces from each plate were cut for physical and chemical tests [6, 4/23/06]. It was well known at the time that the upper portions of an ingot (and thus the upper portions of a rolled plate) were more brittle than the lower portions (and thus had a lower tensile strength), and therefore the top was frequently cut off. What was less well-known was that tensile strength varied throughout the plate due to the movement of the molten steel within the ingot as it was poured. This movement caused bubbles and distributed the chemicals within the steel differently throughout the ingot. Huston had the test plates bisected vertically, planed, photographed, and analyzed for tensile strength. His chemist also took drillings from the cut face and analyzed them for carbon content as well (Figures 21 and 22). Huston then summarized the findings in an article that included both writing and drawing in *Proceedings of the American Society for Testing Materials*:

> The reason that the steel in the ingots varied throughout is that:
>
> As steel cools in the mold a steadily thickening wall of solidified metal forms against the sides of the mold (which is usually of cast-iron) and numbers of gas bubbles form and rise to the surface in the liquid portion, causing a rising current of metal adjacent to the solidified wall with return downward current in the center.
>
> This causes thorough mixing of the portion remaining liquid, but as the carbon and other elements are expelled from the solidifying wall the central liquid portion continually gains in these elements.
>
> This action continues until the temperature of the liquid portion falls to a point where its consistency becomes so thick that the gas bubbles cannot rise through it, when circulation ceases, gas formation ceases and segregation

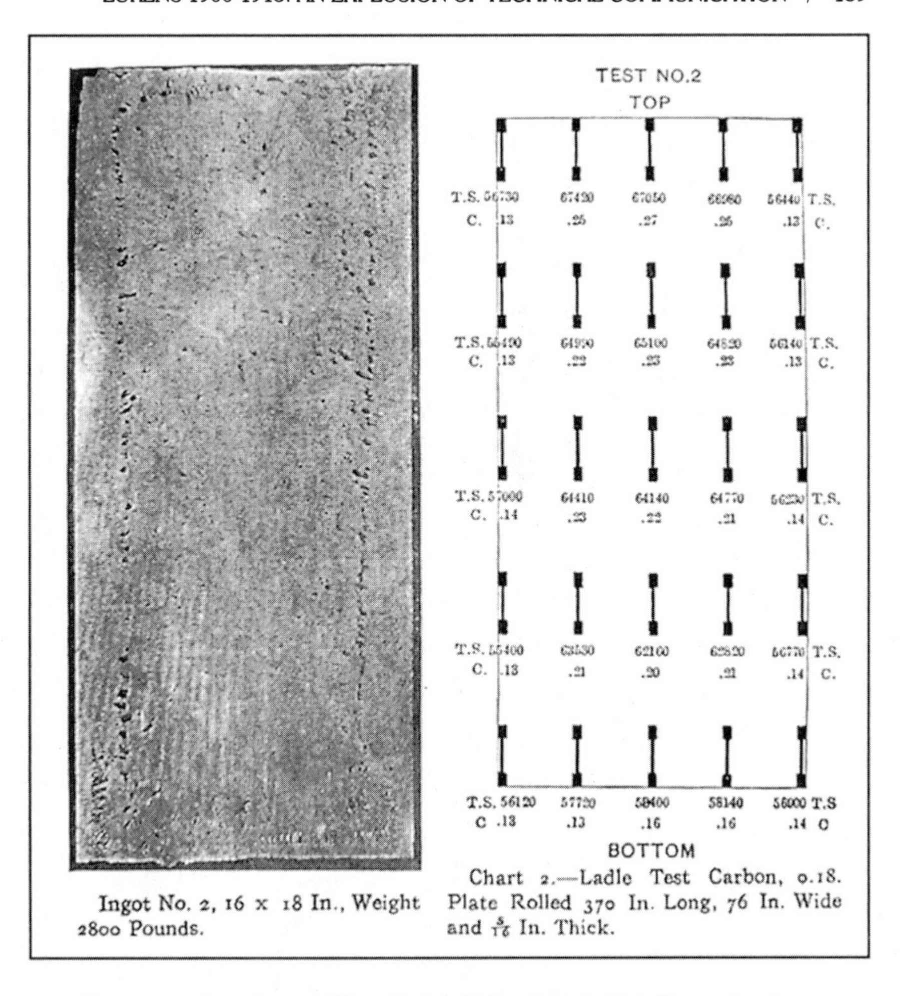

Ingot No. 2, 16 x 18 In., Weight 2800 Pounds.

Chart 2.—Ladle Test Carbon, 0.18. Plate Rolled 370 In. Long, 76 In. Wide and $\frac{5}{16}$ In. Thick.

Figure 21. **Specimen of Steel with Chart (1906).** This illustration from Charles Lukens Huston's article in *Proceedings of the American Society for Testing Materials* shows a sliced ingot with air bubbles and a chart next to it showing the variation in tensile strength after rolling.

> ceases and the metal inside this zone of gas bubbles solidifies in a mass of comparatively uniform character [19, p. 381].

Since many of the test samples used by inspectors came from pieces sheared from the outer margin of the plates, which are naturally softer than the center, the tensile strengths that were stamped on the pieces and used in the inspections did not reflect the plate as a whole. However, the standard practice was that

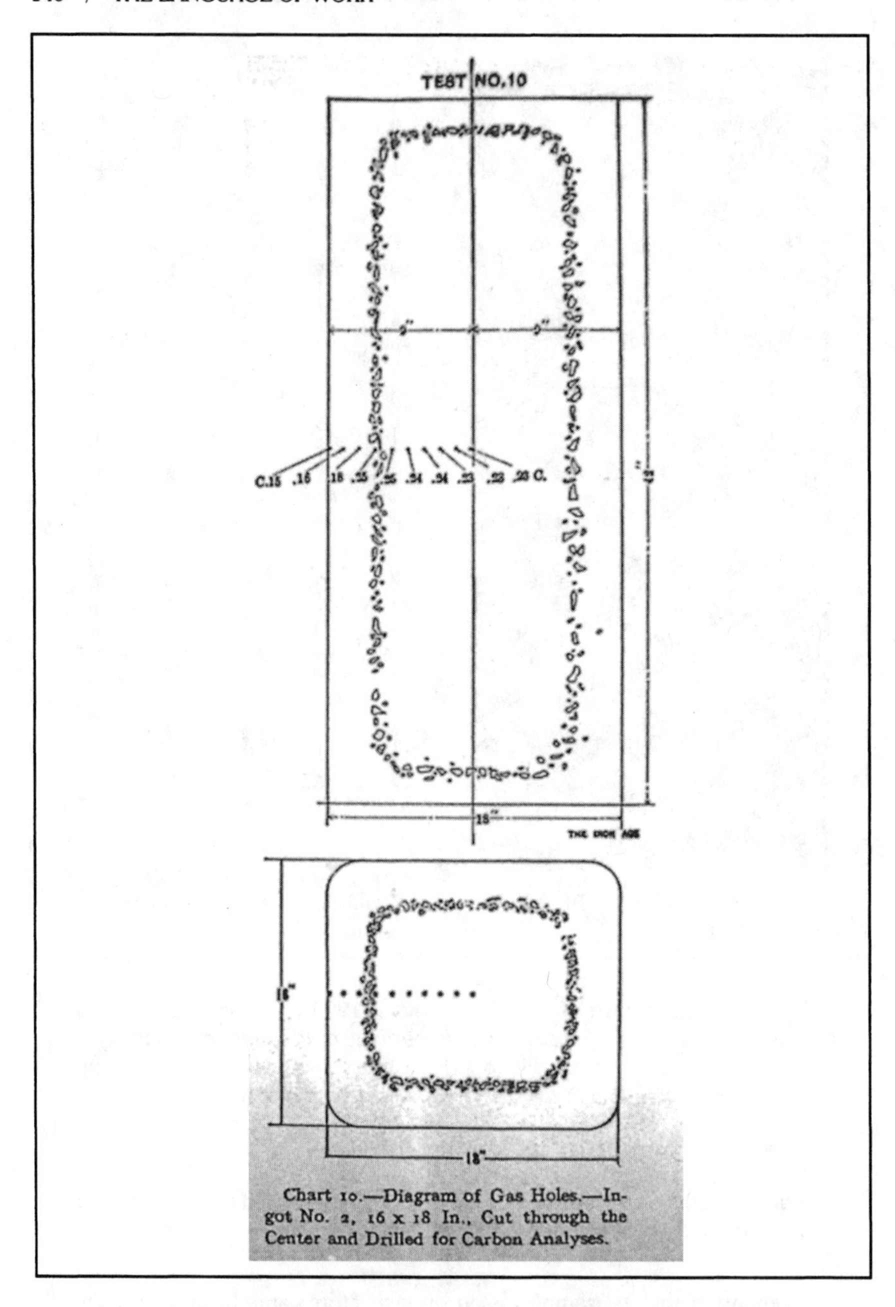

Chart 10.—Diagram of Gas Holes.—Ingot No. 2, 16 x 18 In., Cut through the Center and Drilled for Carbon Analyses.

Figure 22. **Drawing Demonstrating Segregation of Steel in Ingot.** Also from Huston's article, this diagram shows the bubbles with the variation in carbon readings throughout the ingot.

tensile strength was used as a primary indicator of the strength of steel, and the acceptance or rejection of boiler plate depended on it. Huston concluded, "It would be far better I believe if the ductility were introduced as a factor in conjunction with tensile strength to determine the allowable working stresses so that a somewhat lower tensile steel may be used, where it shows a corresponding increase in ductility" [19, p. 383].

This social discourse with the outside world interested in the properties of steel took place within the plant continually as well. At one point Huston advocated stamping the yield point (the point at which a plate would begin to elongate) on the plates as well as the ingot and melt numbers and the tensile (breaking) strength. However, the Engineer of Testing, Howard Taggert, pointed out, "We should not think it advisable to stamp the Yield Point as it would very likely be confused with the T. S. + prove troublesome. As its accurate determination is an impossibility in commercial testing the Tensile Strength, which can be determined accurately, is relied upon to indicate the Yield Point" [6]. Although they knew that the tensile strength was an inadequate measure of plate steel, finding a substitute with which to test was, at this point, a practical impossibility. Charles Lukens Huston continued experimenting with steel throughout the remainder of his career.

The Social Discourse Community to Create National Standards

In 1899 the American Section of the International Association for Testing Materials was formed and began developing standards for steel quality. This social discourse community surrounding testing took place by holding meetings, which were then continued asynchronously by many letters flowing back and forth between the participants. Prior to the start of the ASTM (as it was later called), the setting of standards had been done individually by various agencies, professional associations, and manufacturers; but there was no single entity to collate the standards. The American Society of Civil Engineers, the American Society of Mechanical Engineers, government, and large manufacturers had been setting their own standards. When the American Section of the International Association for Testing Materials first proposed a set of steel standards for everything from railroad track to building beams and wheels to axles, the ASCE and the ASME immediately joined the dialogue, and the resultant social discourse community became a broad, ongoing exchange of information between manufacturers, users, and experts as they negotiated standards and published the results. Charles Lukens Huston served on a number of these committees and contributed to portions of the discourse.

William R. Webster, Consulting and Inspecting Engineer, member of ASCE and ASME, started the American Section of the International Association for Testing Materials. In 1900 he sent out a circular to gather manufacturers to vote on the specifications for "rails, splice bars, structural steel for buildings, structural steel for bridges and ships, O. H. boiler plate and rivet steel, steel castings, steel forgings, steel axles, steel tires and wrought iron" in order to set uniform testing standards [20]. Thirty-four organizations and individuals voted, including the Franklin Institute, Cambria Steel Co., Carnegie Steel Co., Jones & Laughlin, David Thomas, American Iron & Steel Mfg. Co. and National Tube Co. In the first iterations, almost everyone voted yes to everything. Below is one example:

> **Structural Steel for Buildings**
> Steel may be made by either the Open Hearth or Bessemer process.
> Steel shall not contain more than .10% of phosphorus.
> Finished material shall be free from injurious seams, flaws or cracks, and have a workmanlike finish.
> Ultimate Strength of steel 60,000# to 70,000# per square inch. Elastic Limit not less than one-half the Ultimate Strength. Elongation 20%. Bending Test, 180° around a diameter equal to the thickness of piece tested, without fracture on outside of bent portion.
> For the determination of these physical qualities the standard test piece of nominal eight inch gauged length, and the methods prescribed by the Association shall be used.
> Ultimate Strength of rivet steel 48,000# to 58,000# per square inch. Elastic Limit not less then one-half the Ultimate Strength. Elongation 26%. Bending Test, 180° flat on itself, without fracture on outside of bent portion [20].

The American Section of the International Association for Testing Materials, which soon changed its name to the American Society for Testing Materials (ASTM), "quickly gained a reputation as a bastion of manufacturers" [21, p. 236]. No one organization had authority over the creation of standards, and the American Section's presumption of establishing standards made the other organizations angry. The one dissenting voice, voting against accepting the specifications, came from R. W. Hunt, a consulting rail inspector, who wrote, "We are in receipt of your several letters in regard to the vote upon the several specifications for various forms of steel, and also those specifications. The matter has had our most careful consideration, and we must decline to vote in favor of any of them" [20]. Hunt pointed out that the ASTM was largely a group of manufacturers and that specifications had to come from users as well. At the first annual convention of the American Railway Engineers and Maintenance-of-Way Association (AREMWA) in 1900, Hunt stated, "I am a member of an American organization which is seeking to agree upon specifications of all kinds to submit to an International Association as American specifications. I have taken the position that (as unfortunately the organization to which I am alluding is composed almost entirely of the representatives of the manufacturers of the country) they are not the people to put before the world what are thought to be the specifications representing the American ideas and American desires" [20]. Moreover, the ASCE had already come up with their own specifications. However, since the other organizations lacked knowledge about metallurgy, they had to compromise and work with the manufacturers who knew the details of the processes [21]. In order to solve the problems, the various organizations began working together.

Setting standards had been a site of contention and negotiation between manufacturers and users, such as Baldwin Locomotives, Pennsylvania Railroad, and government. Consultants such as the founder of the American Section, Webster, and the rail consulting engineer, Hunt, negotiated to reach an accord, and they succeeded. In the case above, in 1902 AREMWA adopted the ASTM rail specifications (with amendments) and the ASTM adopted AREMWA's specifications for bridges (with amendments). The Master Mechanics Association took up axle specifications, the American Institute of Mining Engineers became responsible for steel forgings and casting, and the American Society of Mechanical Engineers began a committee on steel boiler plate.

Charles Lukens Huston participated on the ASME Board of Boiler Rules (as well as other groups). The members of the Board of Boiler Rules were users, manufacturers, insurance interests, and operating engineers. Howard Taggert assisted by compiling a report on the principal boiler plate specifications used in the United States, listing the name of company, grade, chemical content, and mechanical properties of each. Then the board collaboratively revised a document for the ASME in 1913 [22]. To do this, they created documents by cutting and pasting from the Massachusetts State Board of Boiler Inspection, since it was already largely written (Figure 23). Then they passed the documents back and forth, each participant noting changes in the margins (Figure 24). This document in particular (there were others revised in the same way) was reviewed and revised by Prof. C. F. Miller from MIT, C. L. Huston, H. G. Meinholtz of Heine Safety Boiler Works, Prof. R. C. Carpenter of Cornell and Richard Hammond of the Lake Erie Boiler Works. After Charles Lukens Huston made changes, he sent them to Howard Taggert, the engineer of tests in charge of inspection, for review. Through ongoing meetings, negotiations, and revision, working standards were created and updated by a group discourse process.

PUBLIC RELATIONS DOCUMENTS

The final genre that emerged in this fertile time period was technical advertising brochures and public relations documents. Prior to 1900 Lukens used little advertising or public relations—their product had a secure market and they were well known in the field. Later, they began publishing occasional advertisements in trade journals, informing readers of the unique aspects of their company. For instance, they placed a full-page advertisement in the *Railway Gazette* stating that they were "The First to Make Boiler Plates in America." They also published

PART II. — SECTION 2.

Factors of safety. 7. The lowest factors of safety used for boilers, the shells or drums of which are exposed to the products of combustion and the longitudinal joints of which are of lap-riveted construction, shall be as follows: —

(*a*) Five (5) for boilers not over ten years old.

(*b*) Five and five-tenths (5.5) for boilers over ten and not over fifteen years old.

(*c*) Five and seventy-five hundredths (5.75) for boilers over fifteen and not over twenty years old.

(*d*) Six (6) for boilers over twenty years old.

(*e*) Five (5) for boilers, the longitudinal joints of which are of lap-riveted construction and the shells or drums of which are not exposed to the products of combustion.

SAFETY VALVES. 1. Each boiler shall have not less than two safety valves, one set for the allowed pressure and the other 5 lb. higher. Where more than two valves are used on the same boiler (as in cases of operation at 300 per cent of rating), the additional valve or valves should be set to blow at four or five pounds higher than the first valve or valves which start to blow at the maximum working pressure allowed.

SECTION 2.

Safety valves. 1. Each boiler shall have one (1) or more safety valves.

Size of safety valves, spring-loaded. 2. The minimum size of a direct spring-loaded safety valve shall be governed by the pressure allowed, as stated in the certificate of inspection, and by the grate area of the boiler, subject to the following conditions and as shown by the table in paragraph 3 of this section.

Single boilers, and boilers connected and allowed same pressure. *Condition A.* — A single boiler, or two or more boilers connected to a common steam main and allowed the *same pressure:* the minimum size of safety valves for each boiler shall be governed by the pressure allowed, as stated in the certificate of inspection, and by the grate area of the boiler.

Boilers connected and allowed different pressures. *Condition B.* — When two or more boilers, which are allowed *different pressures*, are connected to a common steam main, the minimum size of each safety valve shall be governed by the pressure allowed, as stated in the certificate of inspection, and by the grate area of the boiler; and all safety valves *shall be set* at a pressure not exceeding the lowest pressure allowed. The aggregate valve area shall not be less than that required for the aggregate grate area, based on the lowest pressure allowed, as shown by the table.

Condition C. — When two or more boilers, which are allowed *different pressures*, are connected to a common steam main, and all safety valves *are not set* at a pressure not exceeding the lowest pressure allowed, the boiler or boilers allowed the lower pressures shall each be protected by a safety valve or valves placed on the connecting pipe to the steam main; the area or combined area of the safety valve or valves placed on the connecting pipe to the steam main shall not be less than the area of the connecting pipe, except when the steam main is smaller than

Figure 23. **"Preliminary report of Committee to formulate standard specifications for the construction of steam boilers," American Society of Mechanical Engineers (1912).** This document was passed back and forth between participants in the creation of a boiler code. Each reviewer made comments from which the final version was created.

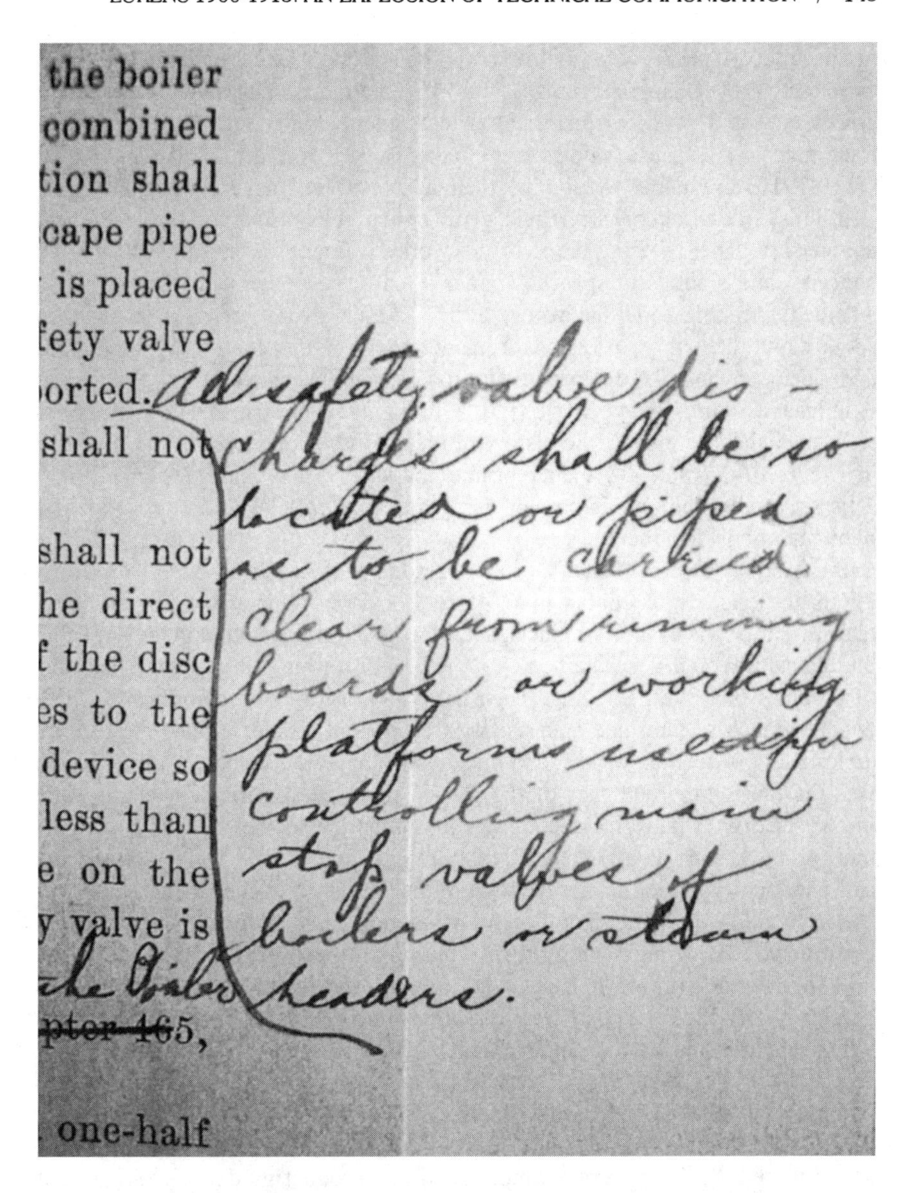

the boiler
combined
tion shall
cape pipe
is placed
fety valve
orted. All safety valve dis-
shall not charges shall be so
located or piped
shall not as to be carried
he direct clear from running
f the disc boards or working
es to the platforms used for
device so controlling main
less than stop valves of
e on the boilers or steam
y valve is headers.
the Boiler
~~pter 165~~,

one-half

Figure 24. **Detail of Report Above (1912).** The participants in this social discourse community made their comments by crossing out sections and writing in the margins.

small cardboard price guides for different products and a longer product guide used by salesmen (see Chap. 6, Figures 13 and 14). Their main outlet for public relations, however, was a series of historical articles about their company in important trade magazines. These articles were accompanied by a centennial celebration (1810-1910) and helped to establish them as a respected name in the industry.

In July 1910 Lukens Steel held a townwide celebration for their 100-year anniversary. There was a parade of the workers, open house at the plant, band concerts, and a series of speeches under a tent. The governor of Pennsylvania attended, speaking about the history of iron in their area; Abram Francis Huston spoke about the history of Lukens Iron and Steel Co.; and John Fritz, the famous and elderly inventor of the three-high process, reminisced about his experiences in industry (Figures 25 and 26) [23]. At approximately the same time, four articles were published describing the history of the company: one in *The Iron Age*, two in *The Iron Trade Review,* and a reprint of the latter in *The Boiler Maker* [23-25]. This set a precedent at Lukens for using historical articles and publications as public relations documents.

In 1912 they also published a product guide that was a small paperback book, 264 pages long, which contained a history of Lukens, a history of making boiler plate, descriptions of their products, and other useful information in the hope that the recipient would keep the book as a reference on his or her desk. The index of products is at the beginning. Then there is a history of the company, which contains much of the same material as in centennial articles. Following that are tables with the heights, widths, and thicknesses they were able to produce, which was the most important part of the guide. The size specification charts were interspersed with narrative sections calling attention to the superior quality of their product, such as a page entitled "Straightening Rolls," which states, "We call particular attention to the fact that all our mills, from the 84-inch up to the 140-inch mill and including the Universal mill, are equipped with specially constructed plate straightening rolls, so placed as to take plates as they leave the mill rolls, while still red hot, changing a wavy and buckled surface into a flat and level one" [26, p. 21]. There are sections on boilers and boiler construction, with useful information relating to chimneys, fuel, heat, water steam, steam pipes, etc. riveting, the strength of rivets, coal, and then a 100-page section called "Sundry Tables of Weights and Dimensions." There were many professionally drawn illustrations that showed specialized products that Lukens hoped to sell (Figure 27). They were attempting, in this product guide, to make a sales document that their readers would keep and use. Lukens was beginning to develop strategies to reach new markets.

* * * * *

The years between 1900 and 1915 saw an increase in the use of technical communication in almost every aspect of the manufacturing process. Many of the

Figure 25. **Centennial Celebration of Lukens Iron and Steel Co. (1910).** After a workman's parade from the factory, the governor of Pennsylvania, John Fritz, and Abram Francis Huston gave speeches.

Figure 26. **John Fritz Delivering his Speech (1910).** John Fritz was a revered and elderly inventor of the three-high rolling mill and former superintendent of Bethlehem Steel.

LUKENS IRON AND STEEL COMPANY

THE LUKENS MANHOLE SADDLE

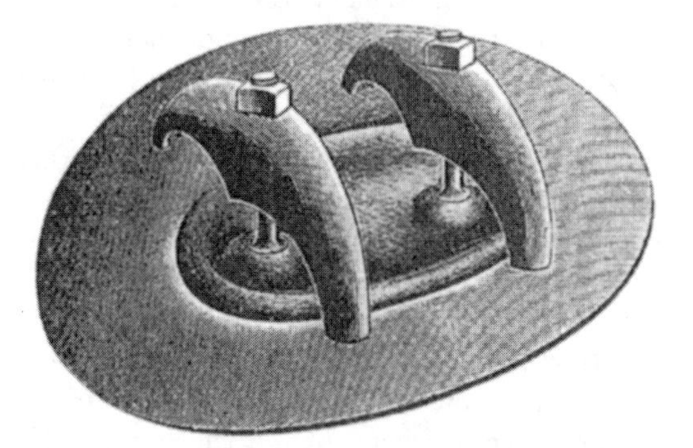

As Applied to Shell of Boiler

Can be set to any diameter of Boiler. Price includes complete fittings plus the cost of the metal in Manhole frame, which weighs about 100 lbs. unless made extra thick.

Figure 27. **Manhole Covers from Product Guide (1912).** This product guide doubled as a handy reference on a variety of subjects, as well as advertising Lukens' products.

old systems continued, but they were constantly augmented by new forms of writing and drawing as multiple people in the plant used these communication methods to define, analyze, and solve problems. The correspondence, previously pressed into letterbooks at a centralized location, spread out into the plant as workers, foremen, and managers used handwritten and, later, typewritten letters and notes to discuss and resolve issues. This new method of using writing and drawing to communicate across the plant began with folded handwritten notes and then grew into perfectly typed complex letters, with multiple copies, made by the stenographer typist. The stenographer typist was the midwife of technical communication, able to take the spoken words of the subject-matter experts, regardless of their level of literacy, and render them clear and understandable. Writing had taken a central role in the organization.

The workers, foremen, and managers at the plant also began to use hundreds of drawings to describe ideas, negotiate new machinery and, finally, as specifications for construction and new machines. Throughout the correspondence (prior to the stenographer typist), many of the letters are written in multiple modes—they contain writing, sketches, calculations, and diagrams. Sketches are appended to documents as well to illustrate an idea. New machinery was built by theorizing mechanical operations and drawing the results on paper, describing them in words, and finally listing specifications for the parts that would be necessary. Many of the machines in the mills had blueprint diagrams that described each part and its interaction with other parts. These diagrams were necessary for planning changes and making repairs. Some material was also ordered by customers in blueprint form. Drawing had taken a central role in the plant.

In the area of testing, the use of writing grew the most rapidly. In the nineteenth century, there was little record of testing procedures, but after 1900 there were in-house testing documents, test documents as part of the manufacturing process, records of scientific tests, published articles of the results, and voluminous records of group efforts to create standards for this and other industries. These testing documents often consisted of drawings, writing, diagrams, graphs, and even microphotographs, which showed the grain of the steel. Hundreds of documents were generated and circulated in the nationwide social discourse community that worked to set standards.

Overall, technical communication went from a peripheral activity in the industry to a central one, necessary for its success. This explosive growth took place in less than 15 years.

REFERENCES

1. A. H. Fay, *A Glossary of the Mining and Mineral Industry*, United States Government Printing Office, Washington, D.C., 1947.

2. J. M. Camp and C. B. Francis, *The Making, Shaping and Treating of Steel* (4th Edition), Carnegie Steel Bureau of Instruction, Pittsburgh, Pennsylvania, 1924.
3. C. T. Baer, *Lukens Steel Company Finding Guide*, Hagley Museum and Library, Wilmington, Delaware, 1994.
4. "206 Mill, 1916-1958," *Lukens Steel Archives*, B-2023, Hagley Museum and Library, Wilmington, Delaware, 1922.
5. J. Yates, *Control Through Communication: The Rise of System in American Management*, Johns Hopkins University Press, Baltimore, Maryland, 1989.
6. "Reports & Memos from Dep't Heads & Workmen," B-4, *Lukens Steel Archives*, Hagley Museum and Library, Wilmington, Delaware, 1900-1914.
7. M. A. L. Aldrich, *New York Natural History Survey 1836-1845*, dissertation, University of Texas at Austin, Austin, Texas, 1974.
8. C. R. Mann, *A Study of Engineering Education Prepared for the Joint Committee on Engineering Education of the National Engineering Societies*, The Carnegie Foundation for the Advancement of Teaching, New York, 1918.
9. E. S. Ferguson, *Engineering and the Mind's Eye*, MIT Press, Cambridge, Massachusetts, 1999.
10. A. F. Hall, "Method of Arranging and Indexing Drawings and Patterns," in *Transactions of the American Society of Mechanical Engineers*, American Society of Mechanical Engineers, New York, pp. 378-387, 1885.
11. J. Brown, Design Plans, Working Drawings, National Styles: Engineering Practice in Great Britain and the United States, 1775–1945, *Technology and Culture, 41*:2, pp. 195-238, 2002.
12. B. Berner, Rationalizing Technical Work: Visions and Realities of the Systematic Drawing Office in Sweden, 1890-1940, *Technology and Culture, 48*:1, pp. 21-42, 2007.
13. B. Rhodes and W. W. Streeter, *Before Photocopying: The Art & History of Mechanical Copying 1780-1938*, Oak Knoll Press & Heraldry Bindery, New Castle, Delaware, 1999.
14. E. L. DiOrio, *Lukens: Remarkable Past—Promising Future*, Lukens' Corporate Affairs Division, Coatesville, Pennsylvania, 1990.
15. *Steam-Engine Indicators and Mechanics of the Steam Engine*, International Textbook Company, Scranton, Pennsylvania, 1906.
16. P. C. Haldeman, "#17—Slab Mill" in "Boiler Tests and Lab Reports 1899-1904," B-310, *Lukens Steel Archives*, Hagley Museum and Library, Wilmington, Delaware, 1903.
17. "Operations Committee Meetings," B-2160, *Lukens Steel Archives*, Hagley Museum and Library, Wilmington, Delaware, 1917-1921.
18. C. L. Huston, Experiments on the Segregation of Steel Ingots in Its Relation to Plate Specifications, in *Proceedings of the American Society for Testing Materials, VI*, pp. 182-198, 1906.
19. C. L. Huston, Practical Experiments in Steel, *Journal of the Franklin Institute, 165*:5, pp. 371-384, 1908.
20. "International Association for Testing Materials: Corresp, Reports, Chargs and Minutes," B-6, *Lukens Steel Archives*, Hagley Museum and Library, Wilmington, Delaware, 1899-1966.

21. S. W. Usselman, *Regulating Railroad Innovation: Business, Technology, and Politics in America, 1840-1920*, Cambridge University Press, Cambridge, Massachusetts, 2002.
22. "Work Copies of [Old] Boiler Code Specifications; Recommendations for Revisions," B-9, *Lukens Steel Archives*, Hagley Museum and Library, Wilmington, Delaware, 1912.
23. "Centennial of the Lukens Iron & Steel Co.: Appropriate Celebration of the Founding of a Great Industry at Coatesville, Pa.—Parade of Workingmen—Historic Addresses by President Huston, John Fritz and Others," *The Iron Age*, reprint (July 7), 1910.
24. "One Hundred and Thirty Years of Iron and Steel Making: Romantic Story of the Development of the Lukens Iron & Steel Co. From a Crude Water Mill and of the Rolling of the First Boiler Plate in the United States," *Iron Trade Review*, reprint (June 30), 1910.
25. "Early Days in the Rolling Mills: John Fritz Tells of the Bitter Opposition of Directors and Employees to His Three-High Mill—Its Successes—How Steel Helped Pay the National Debt," *Iron Trade Review*, reprint (July 14), 1910.
26. *Handbook of Products*, Press of Joseph Glover, Philadelphia, Pennsylvania, 1912.

CHAPTER 6

Lukens 1915-1925: The Union of Words and Work

The evolution of technical communication at Lukens Steel was "additive": when new forms were introduced, they did not replace old ones, but existed side-by-side with them. The account books, the correspondence, the open hearth and defective records, the testing documents and drawing continued, with new forms added as they were needed. As we saw in the last chapter, many more workers in the factory joined the ongoing exchange of ideas, information, and written problem solving in an increasing number of ways. However, the 10 years in this chapter are not significantly different from the previous 15, except for an increasing sophistication in the forms of technical writing and in the addition of another new type of worker, the consultant, who analyzed industrial processes, produced reports, and taught the firm to manage itself with the aid of more written documentation.

Therefore, the form of technical communication that emerged during this time was the report—specifically, the consultant's report. Due to the increasing complexity of the plant and the ever-growing numbers of workers, the plant was no longer manageable by the simple patriarchal system of the past—each unit operated independently, with a loose reporting structure to the top, and no one knew quite how many employees worked there. Also, the salary of many employees was still tied to tonnage produced, so, in quiet economic times, they all quit, waiting for orders to come in to begin work again. To solve the problem, Charles Lukens Huston hired a consultant who taught them to systemize their interactions and document their communications. One result of this was that Lukens held several series of meetings that were transcribed.

Two of the series of meetings were about technical subjects. One was the manufacturing board meeting and the other the operations committee meeting. Both were transcribed by stenographer typists and are clear and easy to read. Thus, we have transcriptions of conversations in which we can "hear" the voices of the plant owners, managers, and workers: the social discourse community was recorded in action. This was one of the major acts of the outside consultant: to institute documentation of the underpinnings and to produce reports that communicated findings to a larger number of people.

THE TECHNOLOGICAL PROCESS: 1915 to 1925

In 1918 Lukens inaugurated the "big mill," also known as "#5." This mill was built in response to the increased demand for ships (boiler plate) during World War I. Unfortunately, as soon as they finished the new mill, the shipping industry in America went into decline, and there was little need for their product. For the first time they had to close part of the plant, lay off workers, and lower wages. The company was in trouble and had to find a way out.

In the years 1915 to 1918, Lukens was very busy with government contracts. In the nineteenth century Rebecca Lukens wrote that they had "all the work we could do," and the same was true during the Civil War and World War I. The exigencies of wartime economy were the only factor limiting their output—it was difficult to get coal and it was very difficult to get labor. Also, in 1918 there was an influenza epidemic that killed 48 workers. Lukens partially overcame the labor problem by hiring black workers from the south and building a camp for single men. Later, they began moving families from the camp into the local housing built for their earlier, mainly immigrant workers [1]. They needed as much labor as they could get, because in 1916 they began their biggest construction project—the big mill (Figure 1).

Plans for the 204-inch four-high reversing mill began in 1916. This mill was designed to roll large marine boiler plates to fulfill wartime government shipping contracts. Charles Lukens Huston designed the four-high system, in which two large supporting rolls gave extra strength to the rolls that came in contact with the steel. His thought process in developing this new type of machinery can be seen in sketches, letters, blueprints, and finally in the mill itself. He worked with United Engineering & Foundry Co. to design the mill and later took a patent out on his innovation. The 204-inch mill was integrated with the 140-inch mill so that ingots could be passed, via transfer tables, between them. Its size dwarfed the original 84-inch mill built in 1870 (compare Figure 2 to Chap. 4, Figure 2, p. 78).

At the same time they were building the big mill, they also built eight 90-ton open hearth furnaces, called Open Hearth #3. However, by 1918 World War I was over, the shipping industry went into a gradual decline, and for Lukens, economic problems began. In 1919 the government cancelled its contracts and in 1920 transportation rates rose due to "Pittsburgh-plus pricing," a monopolistic system created by U.S. Steel that was detrimental to local steelworks. This monopoly did not end until the Federal Trade Commission issued a cease and desist order in 1924 [2, p. 50]. Lukens was no longer able to make a profit on their product, and they shut the new furnaces down shortly after they opened.

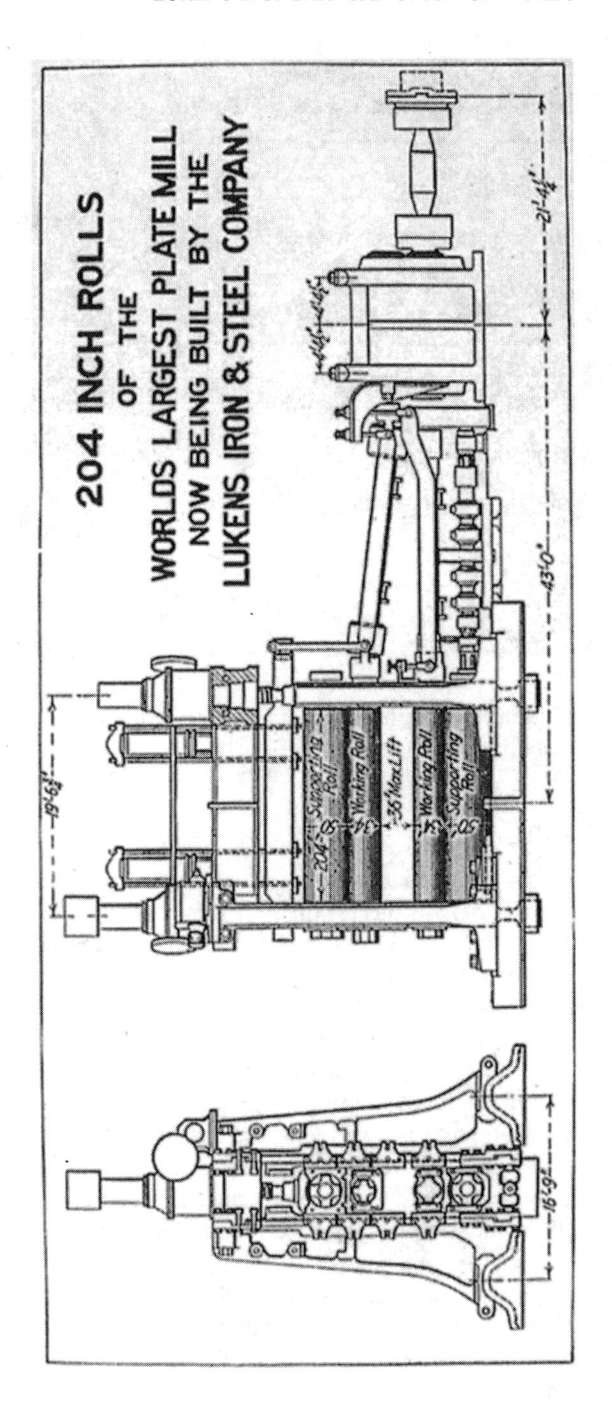

Figure 1. **Drawing of 204-inch Four-high Reversing Mill.** This rolling mill, designed by Charles Lukens Huston with United Engineering & Foundry, was planned in 1916, built in 1917, and rolled its first steel May 22, 1918.

Figure 2. **204-inch Four-high Reversing Mill.** This size of this mill dwarfed the 84-inch mill built in 1870 (Chap. 4, Figure 2). It was capable of rolling steel 25 inches thick and 16 feet wide.

During the early 1920s the mill and its new open hearth furnaces were idle and the stockholders received no dividends [3]. In 1921 they took out a mortgage on the property for the first time: the indebtedness of the firm went from nothing to $5,500,000 [4]. In May 1921, they lowered the worker's wages 16-2/3%. Gradually, they found a market in locomotive "fire boxes," the crown, sides, and combustion chamber that were made from a single piece of steel (Figure 3). They found more business in making foundation plates for skyscrapers [4, p. 3]. They also made "heavy base plates for supporting the columns of the Delaware River bridge, large heavy plates for making glass rolling tables, heavy plates for making fly wheels of the new electrical generators sets, and for the rotors of high speed steam turbines" [5]. Bit by bit they were gaining their business back, but times were hard and emotions ran high as they tried to manage the plant and regain profitably.

One of the problems was that the company had grown haphazardly—each department operated on its own, as an individual fiefdom. There were many departments in Lukens Steel: the open hearths #1 and #2; the connected 84-inch and 112-inch mills; the universal mill; the connected 140-inch and 204-inch mills;

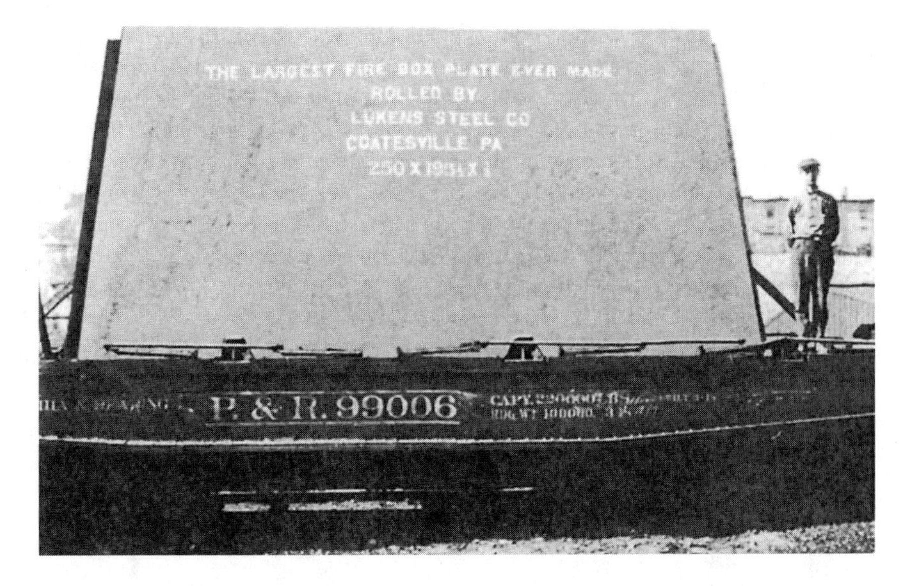

Figure 3. **Fire Box Plate, a Product of the 204-inch Mill.** Ten such plates were provided for the Baldwin Locomotive Works for use in the Atcheon, Topeka & Santa Fe Railroad in 1920.

and many other smaller groups, such as the machine shop, the blacksmith shop, the acetylene area, the foundry, punch shop, grind shop, roll shop, storeroom, pipe shop, pattern shop, tin shop, riggers, carpenters, flangers, locomotive masonry, electrical, chemical laboratory, physical laboratory, commissary, police, material, condenser, track gang, truck gang, general labor, and steam [6]. Each represented a group of men who operated in a slightly different way. For instance, the open hearth and plate mill workers continued to be paid an extra amount by the ton produced. This led to an "elaborate system of dismissing and re-employing men where their work is simply interrupted by lack of continuous mill operations" [7]. The men preferred to work when there were enough orders to make extra money. This created a situation in which they would ask to be dismissed when work was slow and then were reemployed when work came in.

The system of written communication that had developed over the years pertained only to technological developments: Lukens' use of technical communication was entrenched, but there was no parallel system of documentation for managing the plant. The management system that evolved reflected the division between the "works" and the "main office," under the purview, respectively, of Charles Lukens Huston and Abram Francis Huston. There were further divisions between the different shops as they aligned with one or the other. Overall, a major reorganization was required. The simple paternalistic system had become a corporation divided into warring camps between the brothers, between the plate mills and open hearths, and between fiefdoms. The owners didn't have a firm grasp on how many employees they had or how much they were paid. The payroll system was administered by individual foremen handing in sheets listing their workers, who were then paid in cash. Even in the late nineteenth century, when the number of workers had risen to 300 to 400, management of the workers, many of whom were immigrants, was becoming difficult (Figures 4 to 7). During World War I there were approximately 2,000 workers at Lukens and, even after the downturn in the 1920s, the monthly average was 1594 [8].

Charles Lukens Huston was more gifted at understanding the complex interactions in making plate steel than he was in understanding the complex human interactions of the plant. Economic necessity forced the firm to take radical action to systemize the management processes. The office of the comptroller was created and W. J. Bassett was appointed, as comptroller, to be responsible for accounting, cost, time, and payroll [9]. This decision was made by the board of directors and prompted Charles Lukens Huston to resign. Abram Francis Huston resigned as

Figure 4. **Lukens Employees (1895).** These four photographs were taken in the same place on the same day and illustrate the workforce when it was still between 300 to 400. Horace Spackman, who joined the firm as an office boy in 1881 and eventually became vice president, is center front and other managers and foremen are scattered throughout the group.

Figure 5. **No. 1 Steel Plant Work Force.** In the late nineteenth century, the Lukens plant employees were divided into two major groups—the open hearth workers (the "steel plant") and the plate mill workers. Both made extra dividends according to tonnage produced.

well. Abram's son-in-law, Robert Wolcott, took over and Lukens Steel gradually regained profitability. Despite the personnel problems, the ongoing work to manufacture a dependable product continued. During this period correspondence increased exponentially and Lukens Steel was successful in keeping up with the ever-increasing need for complex technical communication in modern industry, even through the difficult periods.

THE EVOLUTION OF ESTABLISHED GENRES OF TECHNICAL COMMUNICATION

Correspondence

Even though many types of technical communication evolved at Lukens Steel, the most prevalent genre remained correspondence. Moreover, the other forms of technical communication were usually accompanied by correspondence. If people are the nodes in the network, correspondence is the vector that connects them, the messaging agent, carrying packets of other information along with it. In the social discourse community of the plant, correspondence was the main

Figure 6. **Plate Mill Work Force.** Since the extra dividends earned for tonnage excluded rejected plates, the plate mill workers often blamed the defects on the open hearth workers (and vice versa).

medium of written communication, and it directed and distributed the other forms. A report was never sent without a letter. A series of defects apparent on a "Report of Plates Rejected" was never sent without a letter. Complaints and their proofs were contained in letters. Long-range discussions about complex technological and metallurgical interactions took place in letters. Drawings were never sent without, at least, a note. Correspondence was both the glue and the avenue of distribution for most of the technical communication at Lukens Steel.

The form of letters was also undergoing change. The constrictions of the letterpress books (letters had to be of a certain size, written with a specific ink, and then taken to a centralized location for pressing) had given way to the brief handwritten correspondence on six-by-nine-inch paper (1904 to 1911), and after that, with the advent of the stenographer typist, letters became formalized. Letters and memoranda, both within the plant and from Charles Lukens Huston to the outside world (it is his filing system from which we take our view), were often long and complex and described technological and physical properties in detail. The letters were still tied to spoken language—they were mostly dictated to stenographers who then typed and prepared multiple copies. The act of dictating is not the same as either the act of writing or of speaking in a conversation; it is halfway between spoken and written language. Charles Lukens Huston dictated

Figure 7. **Colored Trimming Group.** Vince Riley, head trimmer, is next to Horace Spackman at center right. Ultimately, many of the plate defects were blamed on the trimming (shearing) group.

his letters, saying "new paragraph" and speaking punctuation out loud. Dictating to a stenographer meant that the author did not have to prepare the final, however, so it was easy to fill many sheets with material.

Charles Lukens Huston wrote long letters, as did his managers and employees (when they had access to a stenographer typist). Correspondence during this time period had a tendency to be long. Even Huston noticed the problem when he wrote didactically to a manager about effective written communication:

> Where a man's mind is busy on something else, and he picks up a report, it takes him a while to get his attention off of the thing his mind has been upon, and on the thing he is reading and, therefore, the way it catches his eye and the way it gets his attention is important.
>
> I have found in myself a tendency to make long paragraphs, but I find that general advice is it much better to divide up your subject into comparatively short paragraphs, or, a little bit like a landing on the stairway, give a little rest to the mind in passing from one part of the subject to another [10].

Huston seldom followed his own advice, especially when he was arguing; then he often flummoxed the reader by using long, highly technical blocks of text, resulting in four-or-five-page missives. After dictating his thoughts, his secretary, Helen Robertson, edited and typed them, resulting in multiple copies of error-free text that could be sent in multiple directions. Other managers also had stenographer typists, if not permanently, at least temporarily when needed.

Although the overall literacy levels of the employees of Lukens Steel are unknown, one thing is certain, the literacy levels of the stenographer typists were extremely high. In the early twentieth century men could still find gainful employment without a high-school education, and so they did. Women, with fewer employment opportunities, had a greater chance of staying in school. Consequently, female stenographer typists became the mediators of technical communication, bridging the gap between different levels of literacy. Other advances in office technology, especially carbon paper, made larger runs of documents possible. The communications within and between industries increased exponentially and writing became an integral part of the industrial process. The majority of written technical information at Lukens Steel between the years of 1915 and 1925 was correspondence.

The "prediscursive" technical communication discussed earlier in this book—wherein people learned from each other, in person—still went on, but it was augmented by an ability to send a written message to communicate mechanical failures and discuss other issues. The following is an example of a letter written by the master mechanic, who had redesigned an agitator, simplifying it so that it would be independent of skilled labor:

> The arm itself instead of being made of steel casting, I have made from drawn seamless tubing which we have found is the best thing to use for this purpose. We have been using double X heavy pipe but this would scale off and give trouble, so we have adopted the seamless tubing. Where the thread is cut on the tubing we will build up with either acetylene or electric welding in order to get enough to chase the coarse thread. This arm is screwed into a sleeve, as you will note by the print, against a brass collar which will be peened in the steel sleeve in order that it will always be in position when the arms are being changed. The inside piping will telescope into the brass T. This will allow a small amount of water perhaps at first to leak where it is telescoped together, but this will soon plug up and in a very short time leakage will be negligible. The brass collars hold the T and the pipe in position so that one will slide within the other [11].

Although we may not understand this technical explanation, Charles Lukens Huston and others in the plant did. Physical presence, therefore, was no longer necessary to convey the information—it could be done in writing, on paper.

Huston dictated to his secretary who typed and filed the letters alphabetically, starting anew each year. By far the greatest amount of material in the correspondence is about testing. His main correspondence was with the social discourse communities surrounding the creation of standards and studying the basic traits of rolled steel. Much of this correspondence is with various professional associations (especially the ASTM and AMSE Boiler Code Committee); outside experts such as Henry Howe, a professor at the Columbia School of Mines; and government. A large part of his internal correspondence at Lukens Steel was with his testing staff as they performed experiments and communicated results. There is also a great deal of correspondence with his main managers, Engineering & Foundry Co., Inc. (with whom he designed the big mill), and outside consultants and engineers. The correspondence is a vast web with recurring nodes, too complex to physically map. Huston's correspondence files contain many things that are not letters (because they arrived by letter), such as articles, reports, test results, and drawings. The correspondence of Charles Lukens Huston was voluminous and dwarfs other forms of technical communication during this time.

This change—that people were able to write letters by speaking to a stenographer who then prepared final typed copies—is a significant one. Charles Lukens Huston's handwriting is extremely difficult to read. Before the typewriter, he could communicate only to those who had the patience and skill to decipher his notes. After Lukens purchased a typewriter, his ideas could be made clear, but only on a limited basis (limited by the number of clerks in the main office who could type). By 1910 "most U.S. government offices had begun to use carbon paper exclusively for making small numbers of copies, and the private sector was soon to follow. Carbon Paper became an omnipresent part of the office scene for many decades thereafter, until they were eventually overshadowed by the electrostatic copier" [12, p. 128]. This was when Charles Lukens Huston and others had personal typists. From that point on literacy was mediated by stenographers and typists, later called secretaries, who were most likely women. These were the midwives of technical communication.

Record Keeping

The record keeping described in Chapter 4 continued throughout this period as well, although the numbers and types of records they kept increased. Many people, both men and women, both in the plant and in the offices, kept a series of records. The type of records they created evolved with need, so there was no overall system, just a hodgepodge of pieces. Later, when they attempted to order the system, they found the number of reports to be one more complex problem to untangle. Below is a partial list that the production engineer required for the yearly payroll audit:

<u>Cost Reports</u>

Daily reports on furnace rebuilding and any other special jobs on request of Supts.

Weekly report of money expended on repairs in the different departments.

Distribution of expenditures on V and TM jobs, weekly and semi monthly.

Semi monthly report of special costs on various jobs as
requested by Mgr. of Production.
Regular routing cost work on pay roll.

Production Reports
Daily reports to mills in changes of paid weight on plates, account
of rejections or for any other reason which would occasion
a change from the records as turned in by the mill.
Daily reports on tonnage produced in mills.

Weekly Production Reports
Report of ingots rolled on each turn in each mill
Report of hollow tops by carbon
Report of surface defects in U. M. by side furnace and pits
Report of scabs in 140″ by size of ingot
Plate mill production report by turns and combined
Report of O. H. [open hearth] production
Report of flue produced
Report of oil furnace tonnage
Report of O. H. defects

Monthly Production Reports
Report on production both of O. H. and plate mills
Report of pit and ladle scrap
Report of scrap, cinder and scale
Rejection report in connection with plate mills
Report on O. H. defects
Report of Flanging Defects [13].

Lukens also kept various time-keeping records, weekly reports of materials and value, weekly reports of fuel oil used, reports of inbound material, and monthly reports of raw material on hand. The reports from this period were still in bound books or on scraps of paper that originated on the factory floor and then were filed in folders, as illustrated in Chapter 4. Although they were necessary to the running of the plant, the haphazard way they evolved had become a problem that had to be solved before the change to a modern corporate structure in 1925.

Drawings and Blueprints

When Lukens began using stenographer typists in 1912, the drawings within correspondence ceased. From that point on, if there were drawings included in a letter, they were included as an attachment. Drawing and visual representation were still essential, like writing, but gradually they were separated and specialized. The technology that became the main conduit of graphic representations was blueprinting. Previously, many of the drawings were still on sized linen. From this point on, drawings were almost invariably on blueprints.

Nearly every construction project, no matter how small, began with a drawing. The design of machinery was communicated with drawings. Expansions of the plant required multiple drawings. During conversations, ideas were communicated on paper with drawings. Also, for the first time graphs and charts were used to visually communicate overall trends, especially in open hearth and plate production.

Blueprinting made the "exactly repeatable image" possible in the manufacturing environment. William Ivins writes about the impact of the exactly repeatable image in our culture, an impact that often goes unnoticed: "There have been many revolutions in thought and philosophy, in science and religion, but I believe that never in the history of men has there been a more complete revolution than that which has taken place since the middle of the nineteenth century in seeing and visual recording" [14, p. 94]. In the nineteenth century the technology became available to produce large print runs of publications with finely detailed illustrations. These books, journals, and newspapers increased information flow throughout America, making advances in science and technology occur more rapidly. Copying processes such as blueprinting enabled knowledge codification and transfer on a scale that allowed us to rise above the individual memory to the group memory, thereby allowing the creation of more complex objects. All of these things—writing, visual reproduction, technology, and science—evolved together.

Most of the formal drawings were made into multiple copies by the blueprint process. Like correspondence, the number of blueprints grew rapidly; mills, furnaces, machines, buildings, and even railroad track plans had blueprints. Changes were planned with drawings. The example below is a small section of a much larger blueprint showing the layout of the big mill and its machinery (Figure 8).

Drawing was still used to convey ideas on an informal basis as well. There are sketches scattered throughout the company files, and many ideas were communicated, at least partially, by drawing. For instance, a letter from the manager of operations, William Hamilton, to Charles Lukens Huston talks about the proposed changes they were making to a furnace, but then states that since the formal drawings were not yet ready, he would enclose a sketch that illustrated the changes. The sketch was a simple one on a standard letter-size sheet of paper (Figure 9). Attached to that same set of correspondence is another letter-size paper with a sketch of furnace brickwork and numeric calculations, the latter probably from Huston's hand as he calculated the surface area of the brickwork (Figure 10).

They also began to create tables and graphs to visually display production information. "In the early twentieth century, graphs came to be widely accepted as a useful technique for reporting large amounts of data in a readily accessible form" [15, p. 85]. Graphs could communicate trends much more rapidly than lists of numbers. Since the most convenient forms of copying were carbon paper and blueprinting, graphs were sometimes done in blueprint form (Figure 11).

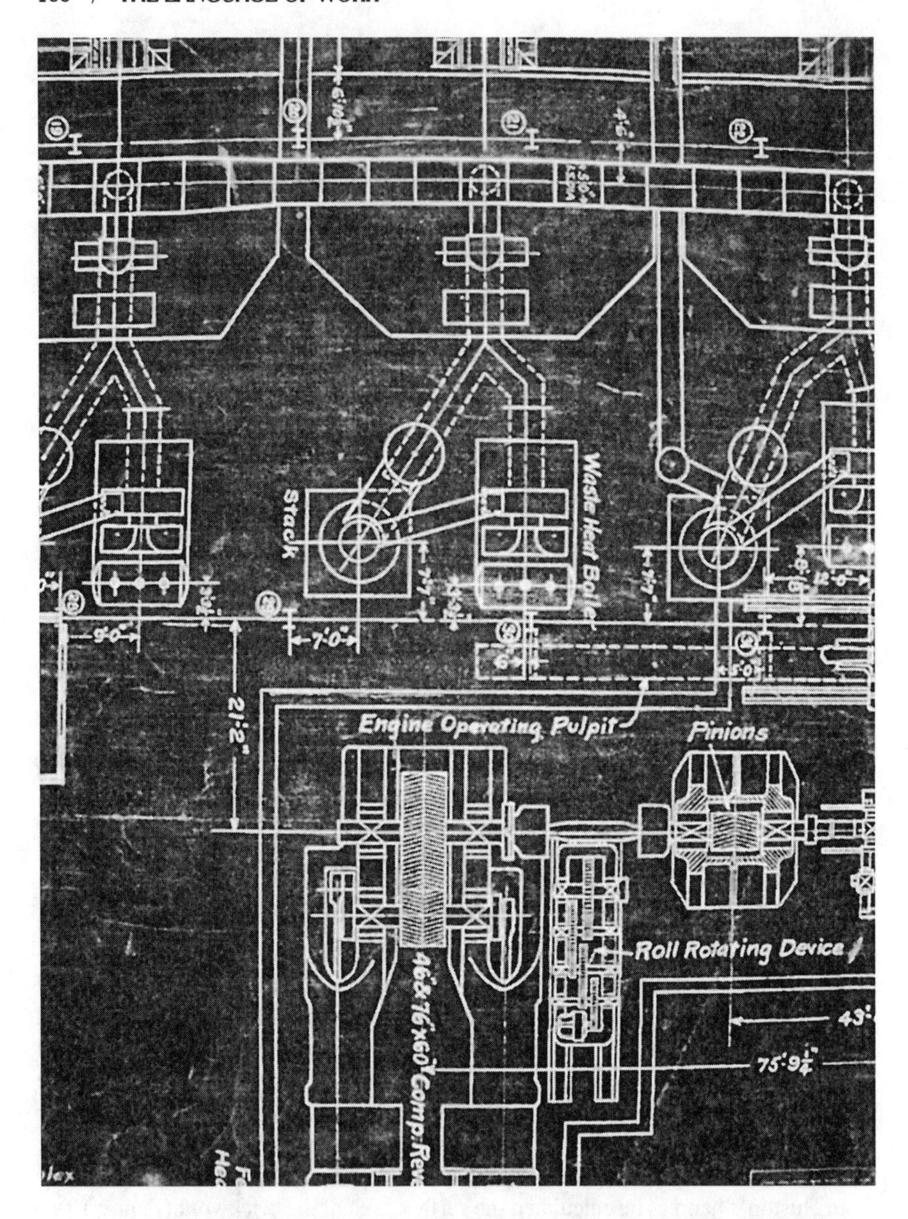

Figure 8. **Detail of the General Arrangement of the #5 Mill, Revised.** This is a small section from a blueprint floor plan, approximately three by five feet, that shows the placement and interaction of the machinery. This blueprint shows frequent use since it is ripped in half and additional revisions were made on fresh paper glued to the original.

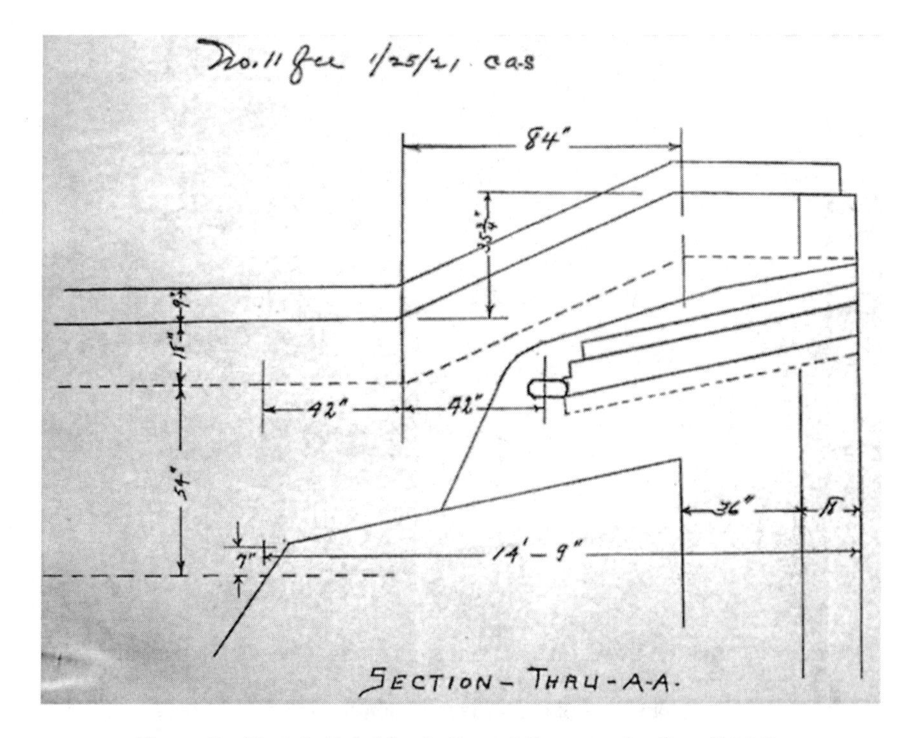

Figure 9. **Sketch Used for Informal Communication (1921).** The manager of operations sent this sketch to illustrate the planned changes in a furnace.

Drawing and blueprinting were fully integrated and important parts of technical communication at Lukens Steel in the early twentieth century. Neither writing nor drawing alone could have conveyed the necessary mechanical and physical information for creating such complex machinery and the equally complex interconnecting parts. Thus, the spoken word (and physical presence) in prediscursive technical communication, in combination with the narrative structure and capacity for containing detail inherent in writing and the abstract representation of spatial and mechanical ideas in drawing worked in concert to convey complex meanings to build a complex new world.

Scientific Testing and Evolving Standards

The act of testing had become as fundamental as the act of technical communication to the manufacturing process and to creating a stable product. There were many variables in the manufacture and rolling of steel, and no one person or group—manufacturers, government inspectors, professional associations, or

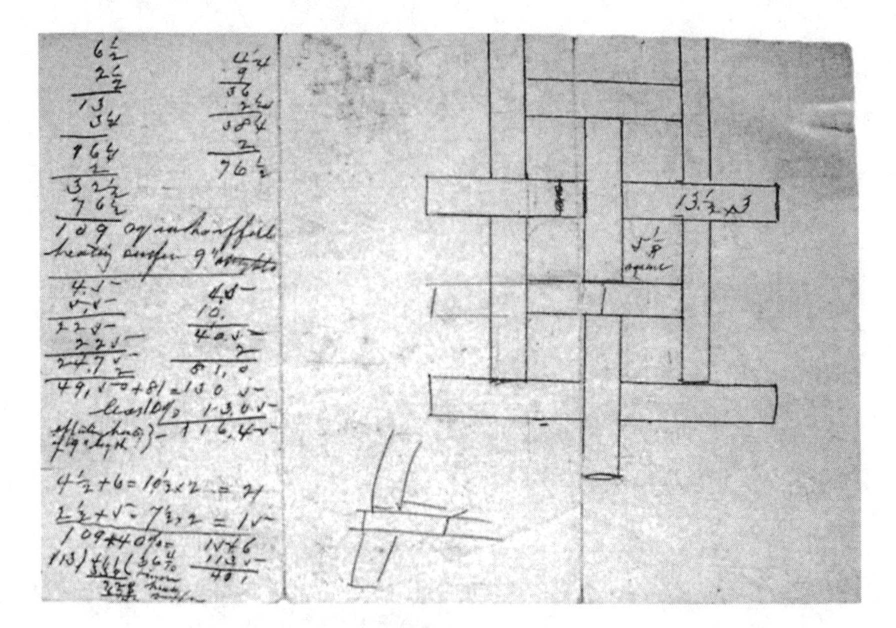

Figure 10. **Sketch Used for Thought Processes.** Charles Lukens Huston made these calculations to check the amount of exposed surface in the brickwork of the furnace structure proposed in the previous sketch (Figure 9).

scientists—had all of the answers. Mostly they had questions, and they tried continually to solve them. Lukens Steel and Charles Lukens Huston participated in this ongoing conversation by offering the factory as a test site, making test samples for others, tracking their own defective plates, revising boiler codes, and experimenting and publishing the results. In a report to the directors of the company, Huston wrote, "It is our policy at all times to keep a certain amount of research work going, for the purpose of checking up the quality of our product, and for being prepared to make special grades of steel . . ." [16]. In a letter regarding rejected plates, he wrote, "There are so many varying conditions to be met with in the service of steel in this respect, that it is difficult to prophesy safely just what the results will be until the conditions are known and tested out." Then he added this joke:

> I remember a story my father, Dr. Huston, told me some years ago of a young Doctor in England, who informed an Irishman, who came into the hospital with Typhoid Fever, that he could not grant his request for salt herring because it would be bad for him. But the Irishman bribed a nurse, got some salt herring and it made him better.

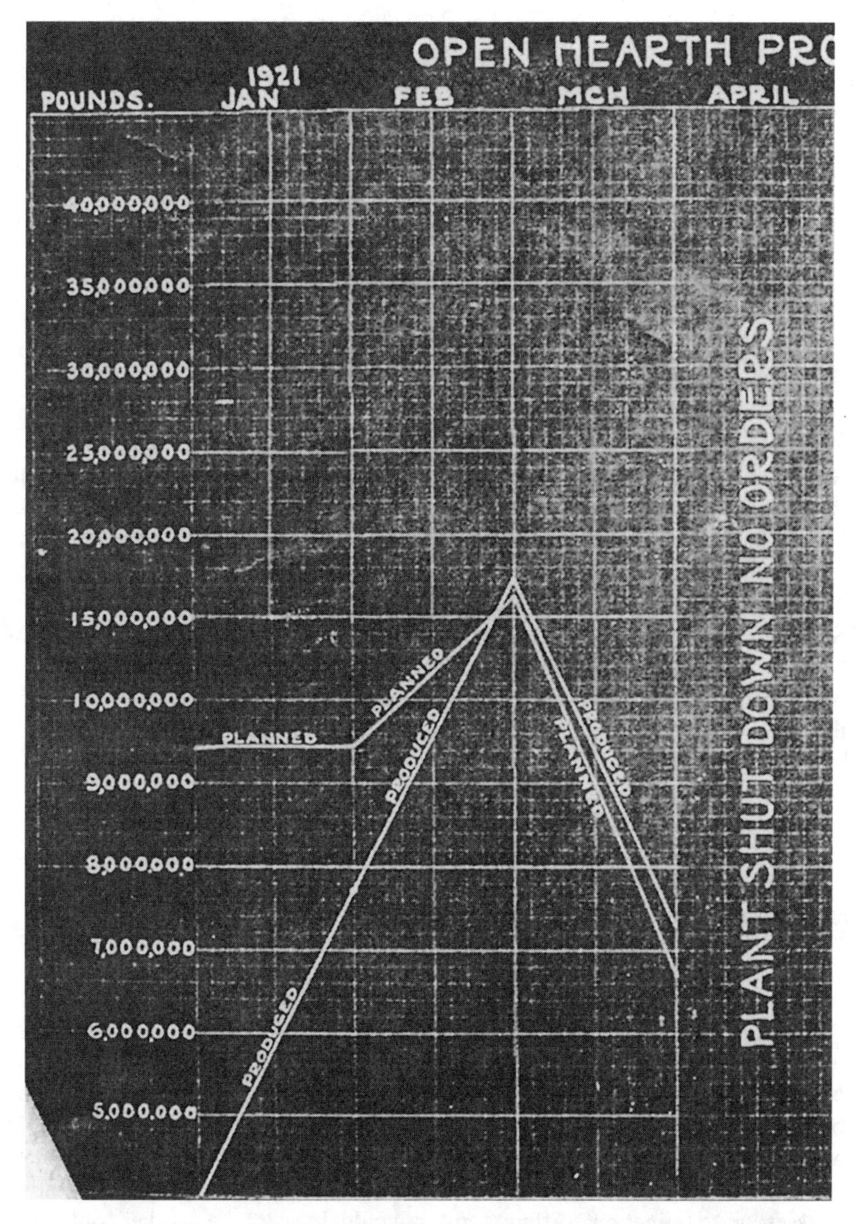

Figure 11. **Blueprint Graph (1921).** In order to make multiple copies of tables and graphs, Lukens used the blueprint process on letter-size paper as in this example.

> The Doctor made a note in his note book that salt herring was good for Typhoid Fever.
>
> Sometime later a Frenchman came in with Typhoid Fever. The Doctor prescribed salt herring for him. The Frenchman died.
>
> The Doctor then amended his notes, saying that salt herring would cure an Irishman with Typhoid Fever, but would kill a Frenchman with the same disease.
>
> The farther we get into this subject [rejected plates] the more we shall know about it [17].

Years of testing were necessary to discover the properties of steel that would best work in marine and other boilers just as years of testing precede, accompany, and follow any new mechanical and chemical manufactured product. Lukens Steel had an advantage in that they were experts at testing and thus able stay at the forefront of the specialty steel market.

The standards for marine boiler plates were an important issue at Lukens Steel, because it directly affected what they could sell. The inspection rules had evolved to a state in which, for heavy boiler plate used in marine boilers, both tensile and bend tests were required [18]. Charles Lukens Huston explained that the earlier tests of tensile strength, alone, were insufficient:

> In the early days of supervision and control by the government, of the materials, construction, operations, etc., of steam boilers on river steamboats, under the United States Board of Supervising Inspectors of Steam Vessels, one of the first rules proposed was that all boiler plate, then generally made of iron, should be required to meet a stipulated minimum tensile strength. Attention of the Board was promptly called to the danger this would involve, by encouraging the use of hard, brittle iron, as the harder the iron the greater the allowable pressure [19].

As iron shifted to steel and boilers became larger, Charles Lukens Huston continued his father's research on the other variables in the melting and rolling of steel that could adequately predict its actions in different environments. In an earlier article he stated, "For material to be subjected to compression or to bending (which is a combination of tension and compression) or to alternating stresses or to vibration, different rules would have to be worked out, each adapting to the conditions of service . . ." [20]. This was of central importance to Lukens Steel because the number of rejected plates was a pressing issue, especially in view of the oncoming Great War.

In 1916 the U.S. Assistant Inspector for the Steamboat Inspection Service changed the rules to read, ". . . you are directed that hereafter all tensile and bending test coupons shall be matched to the plates which they represent, by the inspector, before any test will be made. Furthermore, when plates are intended to be sheared into several patterns they shall not be sheared until after the coupons are matched" [21]. This required an inspector to be present during the manufacturing process so that they could observe the shearing of the testing specimens before rolling could take place, which basically meant that an inspector had to be there all the time. Coupons accompanied the plate throughout the process (Figure 12).

The social discourse community surrounding this issue, composed of government, manufacturers, and associations, continued to negotiate to make the tests as effective as possible without stopping the flow of manufacturing, but it was a difficult problem to solve. In a report from the Bureau of Standards containing the minutes of a conference held in Washington in 1917 with manufacturers and representatives of the United States Shipping Board Emergency Fleet Corporation, the author of the report, S. W. Stratton, summarized:

> The production would be greatly facilitated and rejections reduced if the United States Steamboat Inspection rules were to be amended to specify one quench bend test and one tensile test from each plate. It is recommended that the quench bend test be taken from the part of the plate which represents

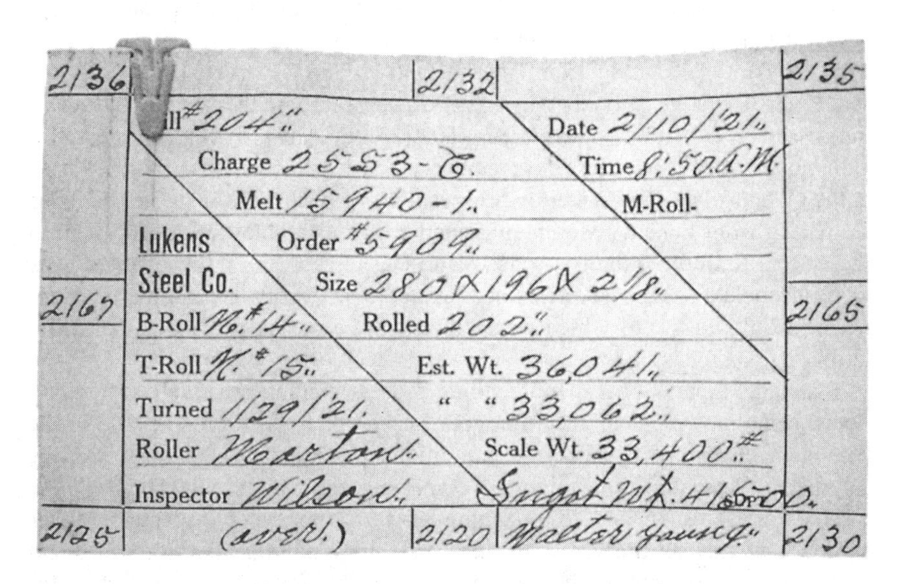

2136 2132 2135
Mill # 204 Date 2/10/21
Charge 2553-G. Time 8:50 A.M.
Melt 15940-1. M-Roll
Lukens Steel Co. Order #5909
Size 280 X 196 X 2 1/8
2167 B-Roll No. #14 Rolled 202 2165
T-Roll N. #15 Est. Wt. 36041
Turned 1/29/21 " " 33,062
Roller Morton Scale Wt. 33,400#
Inspector Wilson Ingot Wt. 41,600
2125 (over.) 2120 Walter Young 2130

Figure 12. **Marine Boiler Plate Coupon (1912).** This coupon, which was attached to each plate, records its history, including the melt and ingot number and the names of the roller, manager, and inspector. These coupons were required for inspection purposes.

> the top of the ingot, and the tensile test from the part which represents the bottom of the ingot. It was pointed out that two tests as specified above are required under the following rules:
>
> American Society of Testing Materials
> Pennsylvania Railroad Company
> New York Central Lines
> American Locomotive Company
> American Bureau of Shipping
> Lloyds (If the plates weigh 5,600 pounds or less) [22, p. 3].

Each of the above entities created their own standard testing procedures for boiler plate. The above report refers specifically to the article published by Charles Lukens Huston in the *Proceedings of the American Society for Testing Materials* in 1906. The report was part of an effort to combine the various standards into one national set. Since the manufacturing process continually changed, the testing procedures had to evolve as well. The government, boiler makers, scientists, and manufacturers all worked together, and documents went back and forth between government officials, manufacturers, committees, and professional associations to vet the changes.

As a manufacturer, Lukens Steel had access to the actual working conditions in which the steel was made, so several other stakeholders in the standards issue used the plant or its product for experimentation. The Department of Commerce sent a representative to Lukens to discuss ideas regarding the specifications and the possibility of testing at the Lukens plant [23]. Dr. Henry Howe, a professor at Columbia College and a leading metallurgist, also planned and performed tests at the Lukens plant. The Watertown Arsenal in Massachusetts requested samples for testing from Lukens Steel in 1915, and a Lukens employee brought them to the arsenal to observe the tests. As was often the case, the result of these tests were inconclusive because the subject matter was so complex. However, these testing results, many of which were published, became a part of the scientific knowledge on the issue.

Testing as a part of the manufacturing process continued as well, generating documents that were then checked against each other and filed. The title of the person in charge of testing as a part of the manufacturing process was the Engineer of Tests, the most important employee regarding testing. The Engineer of Tests was responsible for planning all tests, notifying inspectors (either stationed at Lukens Steel or traveling), making appointments with them, and getting the product out on time. He was responsible for all delays incurred, for meeting the schedules of the sales division, and for all documents pertaining to the tests [24]. The complex administrative position kept the engineer in contact with the sales office, because when plates were rejected, the sales office was affected as well. There were several employees under the Engineer of Tests.

Lukens Steel had a chemical laboratory headed by the Chief Chemist, who was responsible for all metallurgical research and was to submit, in duplicate, all findings, chemical analysis, tests and records. He was responsible for standards in the open hearth department and all tests performed therein, and for reporting any problems [24]. Like the Engineer of Tests, the Chief Chemist had employees under him. Of the two positions, the chemist ran more scientific tests than the engineer in an effort to fully understand the properties of poured ingots and of rolled steel. Charles Lukens Huston paid close attention to both the Engineer of Tests and the Chief Chemist in an effort to untangle the qualities of steel.

What is significant about technical writing and testing is that they are inextricably intertwined: testing would not be possible without the reports and communications it generates. Test results in all fields are communicated in documents, usually containing written and quantitative data, sometimes also containing photographs, graphs, and images. In the sciences, the experimental article evolved into a specific format that made the sharing of ideas more precise and predictable. Although absolute control over the steel making process did not happen during the time period covered in this book, technical communication, like scientific writing, has given us "increasingly immense control of the material world in which we reside" [25, p. 13]. Technical writing has increased our power to physically change the world.

Advertising and Public Relations

Lukens Steel Company did not need an extensive advertising campaign because, as specialty steel makers, they knew their customers and their customers knew them. Their main form of advertising was simple price cards that listed the cost of different shapes and thicknesses of steel (Figures 13 and 14). These cards could be handed or mailed to prospective buyers and would also announce price changes when necessary.

Even so, Lukens frequently published articles about their own history. In 1910, at the company's centennial anniversary, Charles Lukens Huston had several articles published in trade journals. In 1916 he launched a similar public relations campaign about the building of the big mill. In 1925 he published a series of three articles in *Systems, the Magazine of Business* about the history of the company: "132 Years Without Losing a Customer," "Through Two Panics and the Civil War" and "Why We Could 'Carry On' in '93," all themes of survival during difficult times. The articles also included some photographs of the early plant and testing machinery (Figure 15). Lukens knew their history was unique and used it for public relations. This is another example of how their facility with language and communication helped their business survive, by placing it on a solid foundation in a historical context.

Manholes, Fitted Complete

Including flanged and faced off manhole, cover plate, bolts, yokes, and rubber gasket.

In plates less than [illegible]" thick	$15.00
In plates [illegible]" and under [illegible]" thick	16.00
In plates [illegible]" and 1" thick	17.00
In plates 1[illegible]" and under 1[illegible]" thick	18.00

For flanging and facing off manhole openings without fittings charge is made on the basis of $[illegible] for each in the lighter gauges with proportionate extra for heavy gauges. For manhole fittings (cover plate, bolts, yokes and rubber gasket) without flanged manhole the charge is $[illegible] per set.

SIZES

No. 1.	:	9 x 14"	No. 5.	:	12 x 16"
No. 2.	:	10[illegible] x 16"	No. 6.	:	10 x 18"
No. 3.	:	11 x 15"			

Marine Manholes

We make a special fitting for use in Marine Boilers known as Lukens Marine Manhole Door which has the approval of Lloyds Register of Shipping, American Bureau of Shipping and Bureau Veritas. This is made in size 12"x16" only. Special prices apply and will be quoted on application.

Saddle Manholes, Complete

Including saddle plate, flanging and facing off of manhole, cover plate, bolts, yokes, and rubber gasket.

Thickness of Saddle Plate		
[illegible]"		$13.00
[illegible]" to [illegible]" inclusive		20.00
[illegible]" to 1" "		21.00
Over 1"	not less than	22.00

Unusually large or special saddle plates will take special prices.

For saddle manhole flanged and faced off without fittings charge is made on basis of $[illegible] each in the lighter gauges with proportionate extra for heavy gauges. For manhole fittings (cover plate, bolts, yokes and rubber gasket) without the flanged manhole saddle charge is $[illegible] per set.

Hand Holes, Fitted Complete

Including flanged and faced hand hole, cover plate, bolt, yoke and rubber gasket

In plates less than [illegible]" thick, size 4"x6" and 3"x7",	$ 9.00 each
In plates less than [illegible]" thick, size [illegible]"x8" and [illegible]"x10"	10.00 each
In plates less than [illegible]" thick, size [illegible]"x10",	11.00 each
If plates are [illegible]" thick	$1.00 extra
If plates are [illegible]" thick	2.00 extra
If plates are 1" thick	3.00 extra

The charge for hand hole fittings without openings is $2.00 per set; the charge for hand hole opening without fittings is, respectively, $2.00 less than prices listed above.

Tube Holes

We are prepared to cut tube holes. Prices will be quoted upon application.

BRANCH OFFICES

PHILADELPHIA, PA.
NEW YORK, N. Y. BOSTON, MASS.
BALTIMORE, MD.
NEW ORLEANS, LA.

SELLING AGENCIES

J. F. CORLETT & CO.
CLEVELAND, OHIO
CINCINNATI, OHIO

A. M. CASTLE & CO
CHICAGO, ILL. ROCK ISLAND, ILL.
MINNEAPOLIS, MINN.
MILWAUKEE, WIS.
SAN FRANCISCO, CAL.
LOS ANGELES, CAL.

A. M. CASTLE & CO., of Washington
SEATTLE, WASH.
PORTLAND, ORE.

All export material handled by
CONSOLIDATED STEEL CORPORATION
25 Broadway, New York City

PRICE LIST

For Flanging Department Work, Including Flanging, Dishing, Manholes, Fluing, Etc.

World's Largest Plate Mill

ELIMINATE TROUBLESOME SEAMS, RIVETS, ETC., BY DESIGNING FOR

Larger Plates and Heads
Flanged Heads in One Piece Instead of Two

Lukens Steel Company

Main Office and Works
Coatesville, Pa.

Effective June 1, 1921

Figure 13. **Lukens Steel Price Card, Outside (1921).** This was Lukens' main form of advertising. The card supplied all of the necessary information for ordering. The flanged heads and manholes were specially formed products to cover the end of the boiler and provide access to clean it.

First to Make Boiler Plates in America

LUKENS STEEL COMPANY

Steel Plates Sheared and Universal Rolled

Please confer with the Mill for any sizes over or under those shown on this card, also for Irregular and Special Flanging.

Prices for Flanging Heads

PER HEAD

Outside Diameter	3/16 Thick	1/4 Thick	5/16 & 3/8 Thick	7/16 & 1/2 Thick	9/16 & 5/8 Thick	11/16 & 3/4 Thick	13/16 & 7/8 Thick	15/16 & 1 Thick	1 1/16 & 1 1/8 Thick
12" and under 17"	$1.00	$1.00	$1.00	$1.00	$1.00				
17" – 23"	1.00	1.00	1.00	1.00	1.00	$1.50			
23" – 29"	1.00	1.00	1.00	1.00	1.20	1.50	$2.00		
29" – 35"	1.20	1.20	1.10	1.20	1.70	2.30	2.70		
35" – 41"	1.60	1.60	1.40	1.60	2.30	2.90	3.50	$4.20	$5.30
41" – 47"	2.00	2.00	1.80	2.00	3.00	3.70	4.40	5.40	6.70
47" – 53"	3.00	2.70	2.40	2.70	3.70	4.70	5.50	6.80	8.40
53" – 59"	4.00	3.50	3.00	3.50	4.40	6.00	6.90	8.50	10.20
59" – 65"		4.30	3.70	4.30	5.50	7.30	8.40	10.20	12.00
65" – 71"		5.50	4.50	5.50	6.70	8.60	9.90	11.90	13.80
71" – 77"		6.80	5.50	6.80	8.00	9.90	11.50	13.70	15.80
77" – 83"		8.20	6.60	8.20	9.40	11.30	13.30	15.70	18.30
83" – 89"		10.00	8.30	10.00	11.00	12.50	15.00	18.00	21.00
89" – 95"		12.20	10.10	12.20	14.30	16.50	18.20	20.30	24.00
95" – 101"		15.50	12.10	15.30	17.60	20.50	23.60	25.60	29.00
101" – 107"			15.00	19.00	21.00	25.00	27.00	30.00	34.00
107" – 119"			18.00	23.00	26.00	30.00	33.00	36.00	41.00
119" – 127"				27.00	31.00	35.00	39.00	43.00	49.00
127" – 133"				32.00	36.00	41.00	45.00	50.00	57.00
133" – 139"				37.00	41.00	47.00	52.00	57.00	65.00
139" – 147"				44.00	46.00	53.00	60.00	67.00	75.00
147" – 155"				51.00	50.00	59.00	68.00	77.00	86.00
155" – 163"				54.00	54.00	66.00	76.00	86.00	97.00
163" – 179"				65.00	66.00	75.00	85.00	96.00	107.00
179" – 183"				73.00	74.00	84.00	95.00	106.00	117.00
Total Depth	3"	3"	4"	4½"	5	5½"	6"	6½"	7"

Where the total depth, including metal, radius and straight flange exceeds the normal total depth given above, 10 per cent. to be added to list for each additional ½" or fraction thereof.

Prices for Dishing Heads

PER HEAD

These Prices Do Not Include Flanging, Simply Dishing

Outside Diam.	Max. Depth of Dish	3/16 Thick	1/4 Thick	5/16 & 3/8 Thick	7/16 & 1/2 Thick	9/16 & 5/8 Thick	11/16 & 3/4 Thick	13/16 & 7/8 Thick	15/16 & 1 Thick	1 1/16 & 1 1/8 Thick
12"	1½"	$1.50	$1.30	$1.30	$1.30					
14"	2"	1.50	1.30	1.30	1.30					
17"	2¾"	1.50	1.30	1.30	1.30	$1.30				
23"	3¼"	1.80	1.50	1.30	1.50	1.50	$2.00			
29"	4"	2.15	1.80	1.50	1.80	1.90	2.50	$3.00		
35"	4¾"	2.75	2.30	1.90	2.30	2.50	3.00	4.00		
41"	5½"	3.35	2.80	2.30	2.80	3.10	4.00	5.00	$6.00	
47"	6¾"	4.20	3.50	2.80	3.50	4.00	5.00	6.00	7.00	$8.50
53"	7¼"		4.50	3.50	4.50	5.00	7.00	8.50	10.00	12.00
59"	8"		6.50	4.50	6.50	7.00	9.00	11.00	13.00	15.50
65"	9"		8.50	6.00	8.50	9.00	12.00	14.00	16.00	19.00
71"	9½"		10.50	7.50	10.50	11.00	15.00	17.00	19.00	22.50
77"	11¼"		13.00	9.50	13.00	14.00	18.00	20.00	22.00	26.00
83"	12¼"		15.50	12.00	15.50	17.00	21.00	21.00	25.00	29.50
89"	13"		18.00	15.00	18.00	23.00	24.00	27.00	32.00	35.00
95"	14"		22.00	18.00	22.00	24.00	28.00	32.00	36.00	41.00
101"	15"			23.00	27.00	29.00	33.00	37.00	42.00	48.00
107"	16"			27.00	33.00	34.00	38.00	43.00	48.00	55.00
113"	17"			33.00	38.00	40.00	44.00	49.00	54.00	62.00
119"	18"			39.00	46.00	47.00	51.00	56.00	61.00	63.00
125"	19"			45.00	52.00	56.00	60.00	65.00	70.00	75.00

Intermediate sizes and depth of dish varying from above, will be quoted on upon inquiry, but in no case less than price of next larger outside diameter shown.

Prices for Flanging Flue Holes

PER HOLE

These Prices Do Not Apply to Collar Flanges

Inside Diameter.	3/16 to 3/8 Thick	7/16 & 1/2 Thick	9/16 & 5/8 Thick	11/16 & 3/4 Thick	13/16 & 1 Thick
6"-7"-8"-9"-10"-11" 12"-13"-14"-15"-16"	$3.50	$4.50	$6.00	$8.50	$9.00
17"-18"-19"-20"	4.00	5.00	7.00	9.00	11.00
22"-24"	4.50	6.00	8.00	10.00	13.00
26"-28"	5.00	7.00	9.00	12.00	16.00
30"-32"	6.00	8.00	11.00	15.00	19.00
34"-36"	7.00	9.00	13.00	18.00	22.00
38"-40"	8.00	11.00	15.00	21.00	26.00
42"-44"	9.00	13.00	18.00	24.00	30.00
46"-48"	10.00	16.00	21.00	28.00	35.00
50"-52"	12.00	19.00	25.00	33.00	40.00

Apply for prices for larger sizes.

Material punched out of centres to be retained by manufacturer.

Sizes other than those shown to be quoted on, but in no case price to be less than price of next larger inside diameter shown on list. An extra charge may have to be made in case holes are to be flanged in very heavy plates, which would entail extra expense for handling.

Figure 14. **Lukens Steel Price Card, Inside.** This was the most important part of the advertisement since it listed Lukens' products by dimension, thickness, weight, and price.

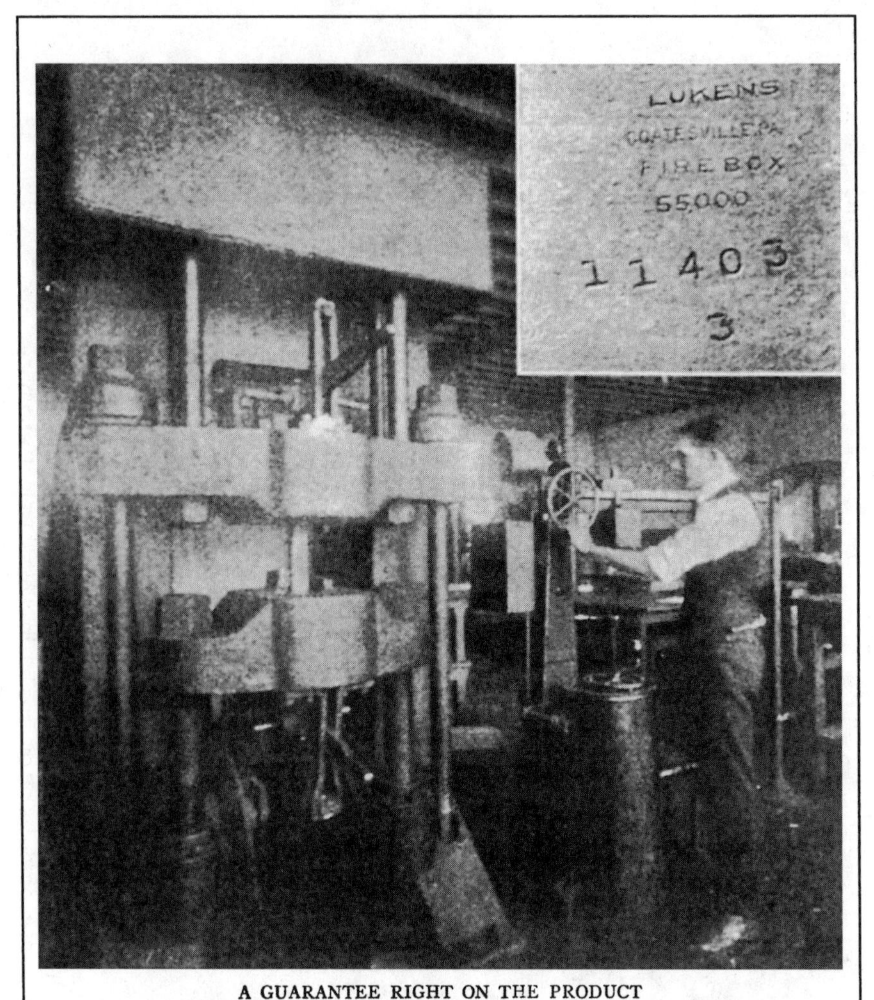

A GUARANTEE RIGHT ON THE PRODUCT

In 1875 Dr. Charles Huston bought a testing machine and began investigating the properties of iron and steel. Today every plate that leaves the factory is stamped to show: manufacturer, location of mill, quality, tensile strength, melt number, and slab number.

Figure 15. **Illustration from *Systems* Article Showing Testing Lab and Caption (1925).** Lukens frequently used their long history, including their early involvement in testing, as a theme for articles describing their business.

THE EMERGENCE OF MODERN MANAGEMENT

In this final section, another new type of worker appeared, the management consultant. Like the field of testing, consulting was dependent on written communication; they, too, were inextricably intertwined. Consultants observed workers and processes and then transferred those observations (and the subsequent recommendations) via written reports. Although not all management consulting writing was a form of technical communication, some of it was. Moreover, like technical communication, it increasingly became necessary to the survival of the company. This story about Lukens Steel cannot end without explaining how Lukens used outside knowledge and outside observers to restructure the firm for the twentieth century. They were still operating in the patriarchal and familial small business owner-operator mode in which foremen and managers reported independently to one of the two brothers, and everyone was part of the family. This was not feasible with 1500 employees.

Although Lukens Steel Company was expert at testing, communications, and making steel plate, they were not innovative managers. As long as their income was larger than expenses, that didn't matter. However, after the end of World War I and the decline of the shipping industry, they lost their major customer base for the new 204-inch mill. In the face of financial problems, they had to face a problem that was social, rather than mechanical or chemical: they had to systemize their plant management. This was made more difficult because of the two warring camps, Charles Lukens (vice president, works manager) and Abram Francis Huston (president, in charge of the main office and sales). When Charles Lukens Huston finally hired a consultant, the consultant only looked at his half of the firm—the plant itself, not the sales force or the extended agent offices in other cities. However, the final recommendations made by the consultant, along with time, forced both brothers to retire and turn the company over to a son-in-law. They never credited the consultant with giving them the knowledge necessary to make this change, but it was the consultant's reports that precipitated the events that led to it. They also tried not to pay the consultant, since consultancy was in its youth and not yet widely accepted in the industrial world.

The company had grown. During the Civil War it employed only 34 men; by the end of the nineteenth century it employed 300 to 400; during World War I, it employed 2,000; but by the 1920s it had dropped to about 1,500. No one knew for sure the exact number of employees and that was one of the problems. Individual departments—each mill, furnace, the flanging department, labor gangs, electricians and riggers, office staffs, and laboratories—had their own system, and each manager continued handing in hours worked by their employees on little slips of paper. Everyone was paid from a cash truck that traveled throughout the plant. The labor gangs were transient and hard to hold, especially during World War I. The number of employees changed every day.

Due to the split between the brothers, the company had no single leader—some reported to one side, some reported to the other. This split was echoed throughout the plant as various workers and managers complained directly to the two heads without following any specific chain of command. Although Charles Lukens Huston was writing articles for *Systems* magazine, there was no system at his plant.

Just like the amount of technical communication expanded, the different forms of accounting writing had expanded. For instance, they had a property appraisal done in 1917, prior to building the big mill, by the American Appraisal Company [26]. In 1924 they hired American Appraisal again to determine the fair market value of the properties as of February 1, 1917, in conjunction with a tax issue [27]. They were attempting to lower their taxes and regain some of the loss incurred from building the big mill #5 and open hearth #3. These documents list every parcel of property, building, and object in it. They also provide maps of the entire plant. In a way this accounting writing is a reification of the plant itself, fixing it in time much as an archaeologist would do in cataloguing the artifacts found at a site. They were also used to try to grasp the whole of the plant in visual and written form, rather than the many parts run by individuals.

Accounting consulting was the first type that they used. Then Charles Lukens hired a new type of consultant—the engineering consultant—after receiving several pitches in the mail offering money-back guarantees. One recommendation from the consultants was that official, transcribed meetings be held. These were meetings of management that were transcribed by Helen Robertson and they are the first time we can "hear" the voices in the plant as they speak. They tell the story of the actual conditions over time. They are silent, however, on the forced restructuring.

Directors and Operations Committee Meeting Transcriptions

Lukens kept official meeting minutes from when they first incorporated in 1899, but they were template descriptions of actions taken. Present at these meetings were A. F. Huston, Charles L. Huston, Benjamin Miller, H. B. Spackman, Howard Taggart and Joseph Humpton. Humpton, the secretary and treasurer, wrote the minutes by hand in a bound book until 1912 when they were typed and placed, with other legal documents, in a binder. These minutes are formalized business procedures recording the closing and opening of the books each year, the reelection of officers, yearly salaries, and the distribution of dividends from profits, which occurred both regularly and on special occasions. If any changes in the plant were undertaken, they were not discussed during the board meetings, they were presented as a *fait accompli*. These minutes continue through 1925, but unfortunately those for the months surrounding the resignation of Charles Lukens Huston and Abram Francis Huston are missing.

After 1917 several new types of regular meetings were held, and transcriptions were taken at the meetings, typed, and bound. The Operations Committee meetings were about the state of the plant and give an overview of the technological and social issues that the company was facing [1]. They took place once per week with approximately 25 attendees between 1917 and 1921. When Huston hired L. V. Estes, Inc., the consultant they sent abolished these large meetings and replaced them with the Manufacturing Board. These meetings took place once per month with approximately six attendees between 1921 and 1925 [28]. Meetings were being systematically used in many corporations by this time. As Yates wrote, "these meetings of foremen and/or middle managers had multiple purposes, including injecting a personal element into work life, promoting esprit de corps, monitoring and comparing the performance of comparable units, discussing policy and generating ideas for promoting efficiency" [15, p. 98]. The meeting transcriptions are interesting in that they are an artifact of spoken discourse: in them you can hear the cadences of the participants, abbreviated by the stenographer typist who retained verbal fragments from the past.

The Operations Committee meetings were attended by the majority of the managers, and each meeting had a general format: W. H. Hamilton started by reading the tonnage numbers for the open hearth and plate mills, and then they began discussing whatever issue was most pressing. In 1918 the most frequently discussed problems were the difficulty getting good coal due to war shortages, labor shortages, and the building of camps for black workers coming from the south. They also discussed general ongoing construction, chemical reactions within the open hearth furnaces, and many other topics such as demurrage, building new tracks, and technical experiments. These transcriptions contain detail that, in combination with the letters from this period, could re-create an exact account of how the company functioned. The language was as close as we can get to the voices of the workers. For instance, the manager of the big mill gave his report:

> Running about fifty-five men on #5 mill. Going to be up against it for cement. Asked A. Goodfellow to take up with H. B. Spackman to get a couple of cars a week. H. B. Spackman said took it up Saturday, but did not know whether there would be any results or not. Ordered all we had of our own thrown in to him. Shall take it up again and see if we can get any result [1, 2/4/18].

The names of the speakers were in front of each transcribed passage, such as in a play script. However, as often happens in conversation, other people broke in and the conversations happened in fragments. Helen Robertson solved this problem in transcription by using the name as an indicator of an issue important to that person. The passage above is attributed to a manager named Dunleavy, and the problem is stated in the first three lines. After he refers to H. B. Spackman, the following line, as indicated by "H. B. Spackman said . . ." was Spackman's response. The line after that, "Ordered all we had of our own thrown in to him" was Charles Lukens Huston, since he was the only one who had the authority to make that statement. "Shall take it up again and see if we can get any result" is Spackman again, since he was in charge of purchasing. The transcriptions are records of real-time verbal interactions at Lukens Steel.

The Manufacturing Board meetings began two months after the last Operations Committee meeting ended in 1920. The smaller meetings were suggested by the management consultant, L. V. Estes, Inc., in order to create a more hierarchical and less lateral system of reporting The only participants were C. L. Huston, H. B. Spackman, W. H. Hamilton, P. R. Baker, and later A. F. Huston and F. H. Gordon, the head of sales. These meetings, therefore, were much shorter than the others. They, too, started with a weekly production report and have many interesting things in them, such as the current economic downturn, the laying off of workers, discontinuing the company camp, the ongoing attempt to get a grasp on how many people were working at the mills and at what cost, and the back-and-forth accusations of who was responsible for rejected plates (furnaces vs. mills). These transcriptions continued through major management changes in the company, and the final meetings in 1925 were led by the new president, Robert W. Wolcott.

In these meeting transcriptions we see the social discourse community together, in one room, interacting in relation to plant processes over time. They show the wide range of topics that they had to cover constantly—chemical, mechanical, and social—and the enormous complexity of running the plant. Increasingly, these problems were solved by using writing in many forms. Miss Robertson recorded that Charles Lukens Huston said, "Mr. Huston would like to have a system started of each man who has charge of a mill making notes from time to time, which will be of value in changes of system, changes of machines, and things that will be of interest in the different departments" [28, 7/16/23]. He practiced what he preached, and many of the employees in the company did as well, leaving a rich written and even verbal record of their technological interactions.

Incorporating Outside Knowledge—Consultant Appraisals and Reports

Lukens Steel has a long history of using outside knowledge within their plant (as well as sharing their knowledge with others). Both management and employees often went on knowledge-gathering trips to see how the manufacturing process was conducted at other plants, and when they returned they wrote reports that were sent to Charles Lukens Huston. When William H. Bischoff was hired as superintendent of the open hearth furnaces, one of his first steps was to go on a research trip to collect knowledge from other companies, much like William Byrd in the 1730s and Robert Erskine in 1770. He went to the Tennessee Coal Iron Railroad Co., the Railway Steel Spring Co., Illinois Steel Co., Otis Steel Co., and Briar Hill Steel Co., reporting the results in a seven-page letter [29]. E. A. Forbes, a long-time employee who replaced Bischoff when he was fired, went with two other employees to the Homestead plant and the Edgar Thompson plant in Pittsburgh, then the Central Iron & Steel Company in Harrisburg and Bethlehem Steel in Steelton, Pennsylvania [30]. Alfred Goodfellow, the mechanical engineer in charge of construction and drawing, went to the Harrisburg Pipe & Pipe Bending Plant in 1921 and the Donner Steel Co. in Buffalo, New York in 1923, writing reports about each [31, 32]. P. C. Haldeman, the master mechanic, visited the Rochester Welding Works, the Birdsboro Works and, when he went to Milwaukee to check on a new gear set that they had ordered, he stopped at Inland Steel Co. and the plant at Gary, Indiana and wrote reports on everything he saw [33]. Moreover, supervising engineers attended meetings of the ASME, the ASTM, the Welding Society, and the Association of American Steel Manufacturers. Lukens Steel knew how to gather information.

Lukens Steel was also adept at incorporating outside information. Charles Lukens Huston was receptive to offers from outside consultants asking to come in to view their company and give an assessment of it. The field of management consulting was new in the early twentieth century [34]. The first consultants sent out circulars and letters advertising their services, often with a free trial period or money-back guarantee. Huston took advantage of these offers. The first consultants that worked for Lukens were Suffern & Son in 1912 [34]. Suffern & Son had begun as CPAs but branched out into engineering consulting. Huston responded to a sales letter that ensured him that an analysis by Suffern & Son could reduce his fuel costs. An engineer was set to analyze the system and issue a report. Huston responded with a five-page, single-spaced letter detailing why each recommendation wouldn't work. He sometimes used this method of overwhelming opponents with technical detail; he was capable of talking as rapidly as Miss Robertson was of typing. At stake, in this case, was a bill for $1,658.57. Suffern & Son replaced the first consulting engineer with a second, who issued a new report, but Huston still refused to pay. The consulting engineer wrote, "The part that really hurts, and it is very hard for me to dismiss from my mind, is the fact that while you claim the services to be of no value, you are still running your producers in accordance with our instructions" [35, 1/28/13]. That consultant was then fired by Suffern & Son.

A variety of other consultants contacted Charles Lukens Huston, but the next agency he used extensively was L. V. Estes, Inc. Estes had a system of frequently sending out sales material such as a booklet called "Human Relations in Industry," accompanied by a letter that offered their services for "the formulation of policies and the creation of an organization for most effectively carrying out the principles outlined" [36]. In technological or testing matters, Lukens Steel was ahead of the rest of the industry. However, in management they were still wed to an antiquated paternalistic system in which two heads officially had authority, but the foremen ruled the floor, and rivalries existed throughout the plant. When the representative of Estes, S. D. Schlaudecker, walked into Lukens, he had no idea of the challenges he was going to face.

According to Christopher McKenna, "Schlaudecker's first step was to draw an organizational chart of the manufacturing division specifying the lines of authority from Huston's position as 'Director of Manufacturing' down to the workers on the shop floor and to have it posted throughout the mill" [34, p. 57]. He found cases of conflicting authority and thus created a structure of reporting that outlined each job exactly. This was not much of a problem—the workers throughout the mill were aware of the structure that led up to Charles Lukens Huston. Communicating along those lines of authority, however, was a different issue. Some workers still went directly to Huston to resolve problems and others went to the managers that they liked best. Early on, Schlaudecker discovered the split in the corporate structure and continually tried to stop the main office, run by Abram Francis Huston, from having free lunches in the company cafeteria, which the plant employees resented. This practice was stopped. Schlaudecker spent time with different managers, analyzing the company and making suggestions, but he couldn't fix the main problem: the rift between Charles Lukens Huston and Abram Francis Huston.

Schlaudecker's first report was October 21, 1920. He sent six more reports during 1920, uncovering more details that should be changed. Charles Lukens Huston started writing long letters disputing various portions of the reports while, at the same time, he implemented other suggestions. On March 1, 1921, Schlaudecker issued an extensive report, summarizing all that had gone before. He covered the social issues: he cautioned Lukens about jealous, derogatory rumors and foremen who were openly antagonistic, and wrote about the rift within the plant management. Schlaudecker also covered technical issues: he made standards for the open hearth department and a weekly graphic chart to aid materials control. He noted that in regard to the physical inventory, however, "no two records agree anywhere." Schlaudecker also had a section called "Excessive Pay Roll." It is this last problem that finally led to the board of director's decision to form the Office of Comptroller and the resignation of the brothers. Although the firm never acted outwardly on the Estes concluding report, the board of directors read it, and it prompted them to take action.

In April 1925 the excessive payroll identified by Schlaudecker took center stage as accountant W. J. Bassett tried to make sense of the finances and personnel issues. Charles Lukens Huston referred to it as "the intricate question of man hours" [37]. Quite simply, no one could figure out how many people worked there. In an article in *Systems*, Huston stated, "We have no formal methods of dealing with the workers" [38, p. 780]. Later that month the Office of Comptroller was formed and Bassett was given the position by a vote from the board of directors. He was appointed authority over the time, wage, and cost departments and immediately announced that he planned to pay the workers by check. Huston objected to this system, saying that it would disturb worker relations, and he resigned. At a meeting of the board of directors (missing from the archives), it was decided that both Charles Lukens Huston and Abram Francis Huston would resign and the company would be headed by Robert Wolcott, Abram Francis Huston's son-in-law. The votes for Wolcott were nine to one. Although the two brothers ran the firm, there was a silent majority in the background who made the decision.

Although Charles Lukens Huston paid some of V. L. Estes's bills in 1920, he stopped paying in 1921, during a year of severe economic hardship. In March Estes himself met with Huston and the board of directors. At the meeting, the board made no final decision on further service from Estes, so Estes remained hopeful and continued billing them. Huston eventually paid most of the bill, but only after long negotiations. In subsequent years Lukens Steel continued to use consultants for research, accounting, and restructuring. In 1927 they hired Ford, Bacon & Davis, who analyzed the plant and made a map with the function of all the buildings identified (Figures 16 and 17).

Consultant's reports are interesting not only because they sometimes contained technical information and laid the structure of an organization bare, but because they did this largely by writing. It is another example of Yates's "rising above the individual memory to a corporate one" except that, in this case, it is rising above a corporation's memory to a professional one, created by consolidating group knowledge from a wide base. Writing had come not only to be used within individual industries, but to connect the industries in a web of shared knowledge, which continued to grow over time. Lukens bought knowledge from Estes just as corporations continue to buy knowledge from consultants today. This knowledge is mainly conveyed in written reports.

* * * * *

Technology became so complex at Lukens Steel that literacy was increasingly a requirement for many of the managers and the foremen. Literacy was spread, also, by the stenographers and typists, who built a bridge between spoken and written language. As the entire works became more literate, the form of communication that became most prevalent remained correspondence. Although we often think of technical communication as manuals, documents, and test reports, letter writing was the most prevalent form (as e-mail is today) and almost everything, including technical material—was distributed by letter.

At the founding of Lukens Steel in 1810, there were one or two people corresponding with the outside world by letter. During the Civil War years there

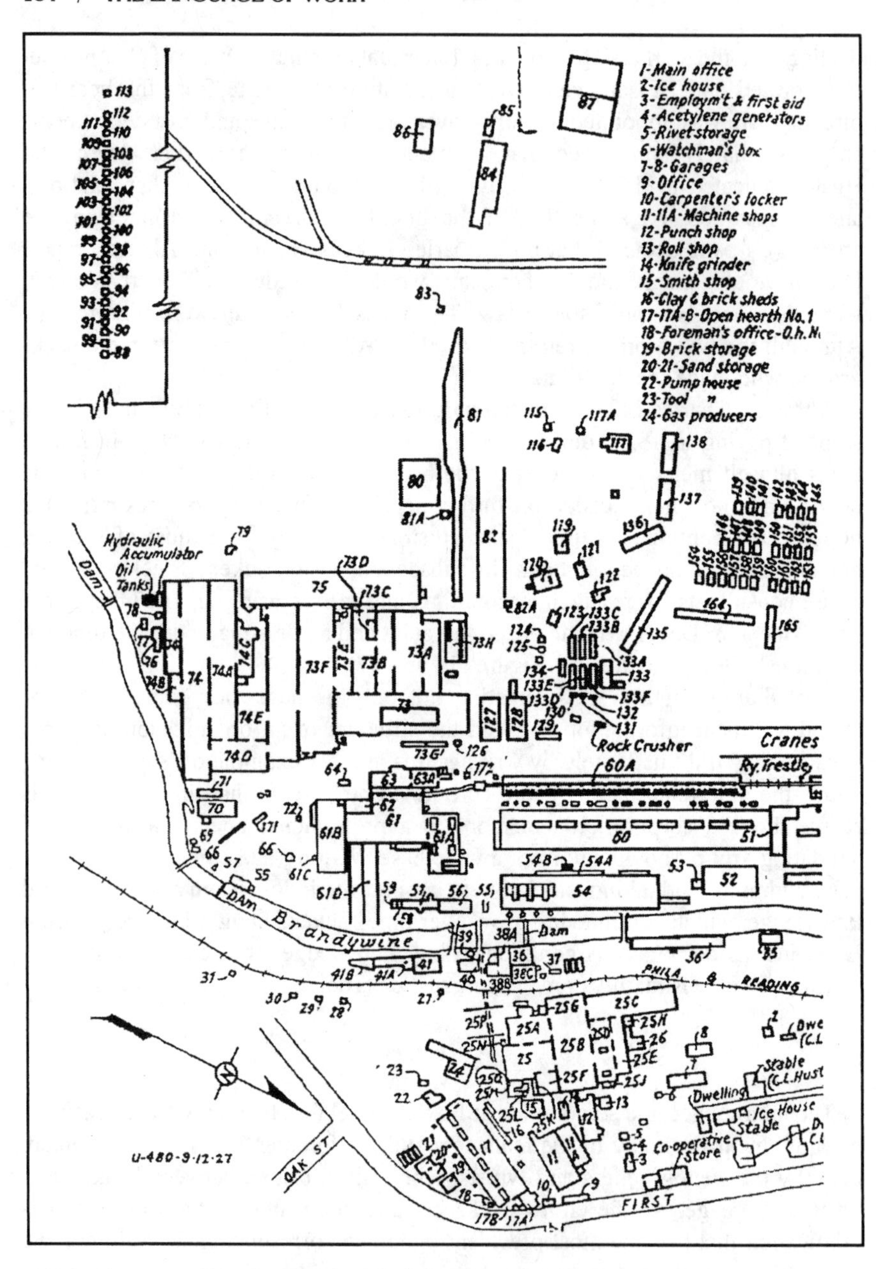

Figure 16. **Ford, Bacon & Davis Map of Lukens Steel Company (1927).** This map is adapted from a small blueprint image. It shows the size and complexity of the plant and identified each building on the site. The earliest portion of the plant is in the lower center.

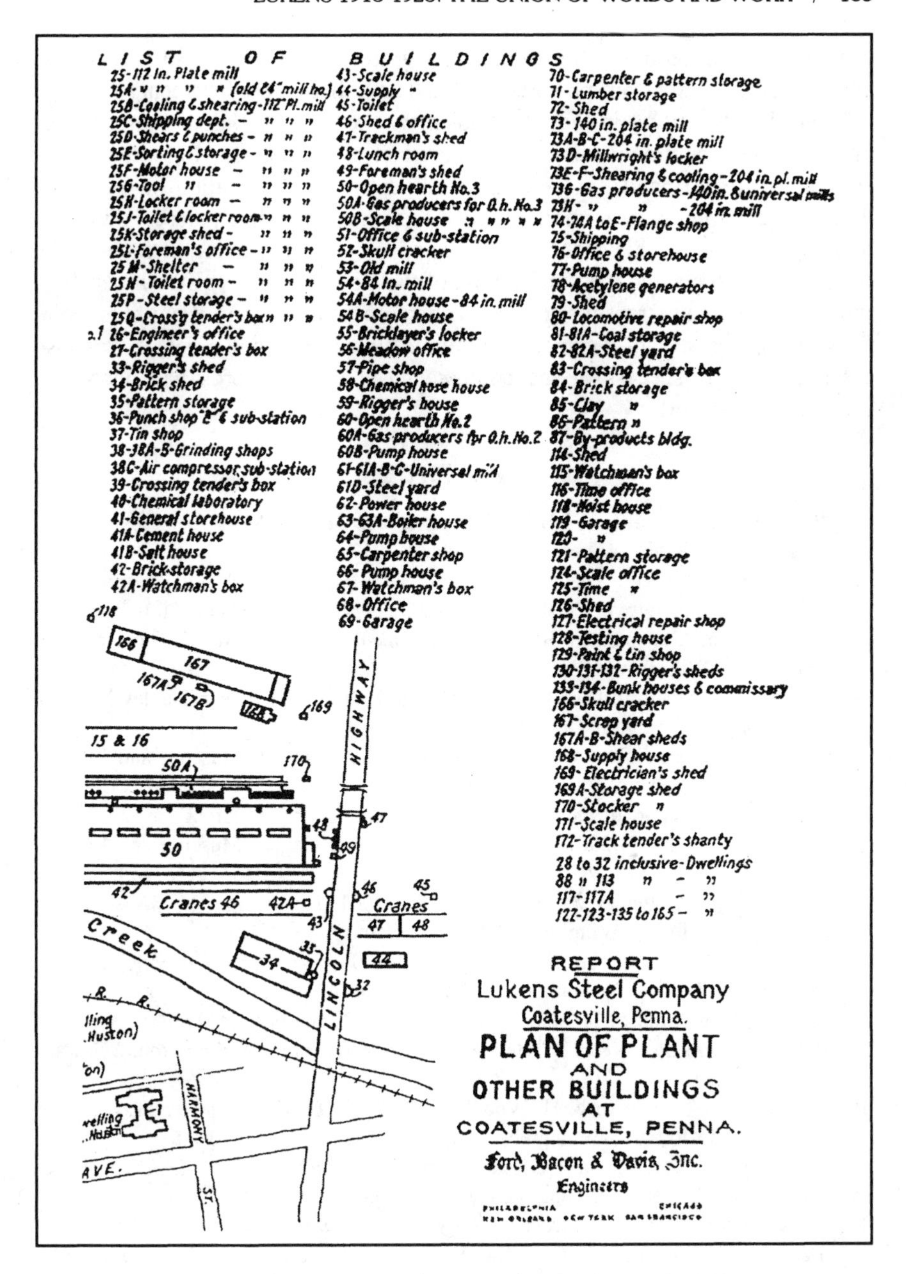

Figure 17. **Ford, Bacon & Davis Map of Lukens Steel Company, continued.** The building attached to the Cooperative Store is the original Brandywine Mansion. Most of these buildings, including the mansion, are intact, providing an example of the industrial archaeology of steel.

were about four people documenting their writing in the letterbooks. After the addition of the first steam-powered mill in 1870, there were between 10 and 12 different hands writing in the letterbooks. By the early twentieth century most managers were communicating within the plant by writing. By 1915 there were so many different hands, voices, and methods of communication in the social discourse community surrounding the production of plate steel, that it is difficult to grasp the whole. This change was accompanied by increasing literacy on the part of the workers and the introduction of the stenographer typist, who enabled rapid and accurate transmission of written communication. From this point forward, the growth in technical communication was exponential. It was also essential for the success of the company. The use of words had become a prerequisite for work.

REFERENCES

1. "Operations Committee, 1917-1920," B-2160, *Lukens Steel Archives*, Hagley Museum and Library, Wilmington, Delaware, 1917-1920.
2. K. Warren, *Big Steel: The First Century of the United States Steel Corporation, 1901-2001*, University of Pittsburgh Press, Pittsburgh, Pennsylvania, 2001.
3. C. L. Huston, "Statement to IRS," B-2002, *Lukens Steel Archives*, Hagley Museum and Library, Wilmington, Delaware, 1922.
4. J. Humpton, "Memo to Directors," B-2001, *Lukens Steel Archives*, Hagley Museum and Library, Wilmington, Delaware, November 1, 1920.
5. C. T. Baer, *Lukens Steel Company Finding Guide*, Hagley Museum and Library, Wilmington, Delaware, 1994.
6. P. R. Baker, "Report of Department Pay Roll and Tonnage Output for Pay Ending January 31, 1924," B-2001, *Lukens Steel Archives*, Hagley Museum and Library, Wilmington, Delaware, 1924.
7. C. L. Huston, "Letter to W. H. Hamilton," B-2001, *Lukens Steel Archives*, Hagley Museum and Library, Wilmington, Delaware, January 19, 1921.
8. C. L. Huston, "Letter from S. N. Dennis," B-2001, *Lukens Steel Archives*, Hagley Museum and Library, Wilmington, Delaware, April 6, 1925.
9. "Extract from Minutes of Meeting of Board of Directors held April 15th, 1925," B-1989, *Lukens Steel Archives*, Hagley Museum and Library, Wilmington, Delaware, April 15, 1925.
10. C. L. Huston, "Letter to F. H. Woodhull," B-2002, *Lukens Steel Archives*, Hagley Museum and Library, Wilmington, Delaware, November 24, 1923.
11. C. L. Huston, "Letter from P. C. Haldeman," B-2001, *Lukens Steel Archives*, Hagley Museum and Library, Wilmington, Delaware, May 24, 1923.
12. B. Rhodes and W. W. Streeter, *Before Photocopying: The Art & History of Mechanical Copying 1780-1938*, Oak Knoll Press & Heraldry Bindery, New Castle, Delaware, 1999.
13. F. H. Woodhull, "Letter to W. J. Bassett," B-2001, *Lukens Steel Archives*, Hagley Museum and Library, Wilmington, Delaware, April 22, 1925.
14. W. M. Ivins, *Prints and Visual Communication*, MIT Press, Cambridge, Massachusetts, 1978.

15. J. Yates, *Control Through Communication: The Rise of System in American Management*, Johns Hopkins University Press, Baltimore, Maryland, 1989.
16. C. L. Huston, "Report to Directors of the Company," B-2002, *Lukens Steel Archives*, Hagley Museum and Library, Wilmington, Delaware, December 12, 1923.
17. C. L. Huston, "Letter to George L. Gordon," B-2001, *Lukens Steel Archives*, Hagley Museum and Library, Wilmington, Delaware, May 7, 1925.
18. S. A. Houghton, "Failures of Heavy Boiler Shell Plates," *The Journal of the Iron and Steel Institute, 92*:2, reprint, *Lukens Steel Archives*, Hagley Museum and Library, Wilmington, Delaware, 1914.
19. C. L. Huston, "Tensile Strength and Ductility of Steel Boiler Plate and the Tetmayer Formula for Figuring Allowable Stresses," *Power, 59*:21, reprint, *Lukens Steel Archives*, Hagley Museum and Library, Wilmington, Delaware, 1924.
20. C. L. Huston, Practical Experiments in Steel, *Journal of the Franklin Institute, 165*:5, pp. 371-384, 1908.
21. E. G. Allen, "Letter to Lukens Iron & Steel Company," B-1988, *Lukens Steel Archives*, Hagley Museum and Library, Wilmington, Delaware, June 6, 1916.
22. S. W. Stratton, "Report Concerning Proposed Alterations of the Steamboat Inspection Service Specifications for Boiler Plate," B-1995, *Lukens Steel Archives*, Hagley Museum and Library, Wilmington, Delaware, June 22, 1917.
23. S. W. Stratton, "Letter to Mr. Chas. L. Huston, V. P.," B-1995, *Lukens Steel Archives*, Hagley Museum and Library, Wilmington, Delaware, February 24, 1917.
24. L. V. Estes, "Estes Report on Lukens Steel," B-1991, *Lukens Steel Archives*, Hagley Museum and Library, Wilmington, Delaware, 1921.
25. C. Bazerman, *Shaping Written Knowledge: The Genre and Activity of the Experimental Article in Science*, University of Wisconsin Press, Madison, Wisconsin 1988.
26. American Appraisal Company, "Physical Inventory Book for Evaluation Purposes," V-190B, *Lukens Steel Archives*, Hagley Museum and Library, Wilmington, Delaware, 1917.
27. American Appraisal Company, "Report, Lukens Steel Company," V-190C, *Lukens Steel Archives*, Hagley Museum and Library, Wilmington, Delaware, 1924.
28. "Manufacturing Board," B-2158, *Lukens Steel Archives*, Hagley Museum and Library, Wilmington, Delaware, 1920-1925.
29. W. Bischoff, "Letter to Charles Lukens Huston," B-2001, *Lukens Steel Archives*, Hagley Museum and Library, Wilmington, Delaware, October 4, 1920.
30. G. A. Forbes, "Letter to P. R. Baker," B-2001, *Lukens Steel Archives*, Hagley Museum and Library, Wilmington, Delaware, April 13, 1921.
31. A. Goodfellow, "Visit to Harrisburg Pipe & Pipe Bending Plant," B-2001, *Lukens Steel Archives*, Hagley Museum and Library, Wilmington, Delaware, January 21, 1921.
32. A. Goodfellow, "Visit to Donner Steel Co.," B-2001, *Lukens Steel Archives*, Hagley Museum and Library, Wilmington, Delaware, February 24, 1923.
33. P. C. Haldeman, "Letters to P. R. Baker," B-2001, *Lukens Steel Archives*, Hagley Museum and Library, Wilmington, Delaware, 1921-1925.
34. C. D. McKenna, *The World's Newest Profession: Management Consulting in the Twentieth Century*, Cambridge University Press, New York, 2006.

35. "Suffern & Son Accountants," B-1999, *Lukens Steel Archives*, Hagley Museum and Library, Wilmington, Delaware, 1911-1914.
36. "L. V. Estes Inc. Efficiency and Management Studies," B-1991, *Lukens Steel Archives*, Hagley Museum and Library, Wilmington, Delaware, 1919-1921.
37. C. L. Huston, "Letter to W. H. Bassett," B-2001, *Lukens Steel Archives*, Hagley Museum and Library, Wilmington, Delaware, April 23, 1925.
38. C. L. Huston, "Why We Could 'Carry On' in '93," *Systems*, June 1925.

Conclusion

The structure of this book is based on the existing documents in the archives. Therefore, the types of technical communication that I have described are specific to Lukens Steel, not to industry as a whole. Other companies and other industries will have different stories. This analysis is based on an overview of an archive—I read and photographed the documents and then sorted them into groups by similarities. I then separated the material into time periods by major changes that occurred. The emerging types of communication then became the outline of each chapter. In this analysis we can see that the first writing was accounting, that quantitative record keeping came next, and that technical writing gradually spread throughout the plant after those two methods were securely in place. This story ends with the emergence of modern management reporting systems. The one constant is growth: once a new form of communication emerged, it stayed and was joined by other forms.

The pattern that we can see in relation to social discourse communities is that, over time, the community as a whole shifted from prediscursive (spoken) communication to chirographic (written and drawn) communication. These discourses did not produce what Foucault calls a monument, but instead they produced material objects that we use in daily life. The discourses existed for the purposes of knowledge generation, transfer, and storage. They existed in order to solve problems. Unlike prose essays, fiction, or poetry, they were not created by an individual expressing inner realities, they were created by groups mediating collaborative realities. They were written as part of everyday living. Foremen had to keep track of the movements of railroad cars and supplies; mechanics had to explain why a machine broke and how it could be fixed; puddle-mill managers had to list who worked when and what they produced. The works manager, Charles Lukens Huston, had to communicate with the outside world about specifications for resilient steel, the open hearth men had to provide drawings and accounts of the construction and production of their furnaces, knowledge that was gained by visits to other plants had to be shared in writing, and eventually, the stenographer typist took over the mediation of this communication. It was a group product, a group activity, and it had real effects within the world. It produced things that could be touched, inhaled, heard, and seen. It is so integrated within our society that we can barely see it separate from the working world.

In the evolution of technical communication at Lukens Steel, we can see how literacy became an important asset in the workplace in the early twentieth century. Prior to 1900 little technical communication took place; after 1900 it was a requirement for any managerial position. A new type of worker appeared early in the twentieth century, the stenographer typist, who could take the spoken language of experts and make it accessible across time and space. Later another new type of worker emerged, the consultant, who was able to analyze and restructure organizations and did so largely through the medium of writing. Writing became an essential part of the industrial process.

We often think of technical communication as dusty documents that sit, unread, on the shelf. In reality, it is all around us. It is so ordinary that, like air and water, we seldom appreciate its value. It has become a part of the way that we interact with the world. The social discourse community that emerged at Lukens Steel is only one example that has emerged within and between every organization in the industrialized world. To create complex products we need to read, write, and communicate constantly. Thus, the study of technical communication can illuminate areas of the world that are still hidden by the camouflage of the commonplace. By studying these areas we can take more responsibility for our engagement with our world as it actually is, rather than as it should be.

Glossary

Note: iron and steel terminology changed according to location, subindustry, and era.

ASME: American Society of Mechanical Engineers, established in 1880 and still active today.

ASTM: originally (1898) the American Section of the International Association for Testing Materials, later (1902) American Society for Testing Materials; an association of manufacturers, users, and government to collaborate to define standards, still active today.

Bar Iron: a bar of wrought iron, semifinished for further working.

Bend Test: a test to record the ductility of steel. After bending to a specified radius, the surface is examined for cracks.

Billet: a small bar of wrought iron or steel, semifinished for further working.

Bloom: a larger bar of wrought iron or steel, semifinished for further working.

Boiler Plate: rolled iron or steel used in the construction of steam boilers.

Dies: a piece of hard iron, placed in a mortar, to receive the blow of a stamp. Dies are used to shape a wide range of metal tools.

Forge: a generic term for either the equipment or a place to shape hot metal; often an open fireplace with forced draft for heating iron (or a building containing several such fireplaces).

Flanging: making a rim, edge, or lip projecting from plate iron or steel.

Furnace, Puddling: a reverberatory furnace for pig iron in which the fuel is separated from the metal and the iron is "stirred" until it takes on the desired consistency.

Furnace, Open Hearth: a process in which iron, ore, and scrap are melted, with the addition of chemicals, in a large reverberatory furnace, producing steel poured into molds.

Furnace, Pit: a reverberatory furnace under floor level in which ingots from the open hearth are placed for even cooling.

Furnace, Reheating: in the early and mid-nineteenth century, a reverberatory furnace in which puddled bars are piled and reheated preparatory to rolling. In the late nineteenth and early twentieth century, a reverberatory furnace in which ingots from the open hearth furnace are reheated prior to rolling.

Furnace, Reverberatory: a furnace in which the ore does not come into contact with fuel (the flame passes over it).

Heads: flanged end-covers used in the construction of boilers.

Heat: a single batch of open hearth steel. Four to six heats were possible within a 24-hour period.

Indicator Cards: a card with a diagram for recording data about the varying pressure of steam (and thus power) in the cylinder of an engine during a stroke.

Ingot: a block of malleable steel made by pouring from the Bessemer or open hearth steel into a mold.

Pig Iron: crude cast iron from a blast furnace, so called because iron runs down a central trench (sow) into perpendicular pigs.

Manhead: flanged cover for a manhole in a boiler.

Manhole: in a boiler, an opening for men to enter.

Mill: generic term for an entire rolling mill or its parts.

Mill, Finishing: a mill for the final stages of rolling.

Mill, Four-High: a rolling mill with four rolls, two that come in contact with the steel and two to support them, made for rolling heavy thick plate.

Mill, Puddling: see Furnace, Puddling.

Mill, Reversing: a two-high mill in which the rolls can move in the same direction or in opposite directions to pull the plate through and/or send it back.

Mill, Rolling: a generic term for an entire rolling mill or rolling mill machinery.

Mill, Roughing: a mill for the first stage of reducing the size of the bloom or billet for further rolling.

Mill, Slitting: an early mill with a rotating shear to cut thin plate into strips for such products as nails.

Mill, Three-High: three rolls arranged one above the other so that the sheet can pass through in both directions without handing it from the catchers to the roller's side.

Mill, Two-High: earliest form of rolling mill, with rolls that move at the same speed in opposite directions to help pull the iron through.

Mill, Universal: a rolling mill with vertical, as well as horizontal rolls, to shape the edge of the steel.

Nail Rod: iron plate that has been rolled to the appropriate thickness and cut in a slitting mill, prior to being formed into nails.

Open Hearth: see Furnace, Open Hearth.

Patterns: a wooden model of machine parts used to form a sand mold for casting.

Plate: rolled iron or steel.

Raceway: a man-made waterway, usually leading from a dam to a power a waterwheel.

Rolling: shaping iron or steel by passing between two rolls.

Rolls: most often made of cast iron or steel, rolls consist of a body, neck (which rests on chocks) and wobblers (with notches for gears to drive the rolls).

Segregation: the areas of different chemical composition in steel caused by currents in the cooling ingot.

Slab: semifinished rolled steel, wider than it is thick.

Soaking Pit: See Furnace, Pit.

Steel: an alloy of iron and carbon. There are many types of steel with diverse properties that are used for different purposes.

Tensile Strength: a measure of the breaking strength of metal.

Wrought Iron: malleable iron produced by hammering or rolling pig iron, leaving some slag in it, and thus giving it a fibrous structure.

Index

Accounting writing, 60, 63-66, 178
Advertising, 143, 146, 149, 173-175
Agents, sales, 45, 68-69
Agricola, 16
Al-Kindi, Abu Yusuf Ya'qub, 15
Aldrich, Mark, 93
Alexander, J. H., 25
American Appraisal Company, 178
American Institute of Mining Engineers (AIME), 27, 30, 143
American Iron & Steel Mfg. Co., 142
American Iron Association (also American Iron and Steel Association; American Iron and Steel Institute), 27
American Locomotive Co., 134
American Railroad Journal, 31
American Railway Engineers and Maintenance-of-Way Association (AREMWA), 142-143
American Society of Civil Engineers, 141-143
American Society for Testing Materials (ASTM), 27, 134, 138-143, 163, 181, 191
American Society of Mechanical Engineers (ASME), 27, 116, 141-145, 163, 181, 191
Arabic texts, 15, 16
ArcelorMittal, 54
Archaeology, 2-5, 8, 178, 185
Association of American Steel Manufacturers, 181
Atcheon, Topeka & Santa Fe Railroad, 157

Baird, Henry Carey, 25
Baer, Christopher T., 56
Baker, P. R., 180
Baldwin Locomotive Works, 45, 134-135, 137, 157
Barnes, E., 122, 124, 138
Baltimore Iron Works, 18
Bar Iron, 39, 44, 59, 191
Bassett, W. J., 158, 183
Bazerman, Charles, 8-9
Bend test, 134, 170-171
Bethlehem Steel, 54, 148, 181
Billets, 47, 61, 76, 191
Biringuccio, Vannoccio, 16, 27
Bischoff, William H., 181
Blueprinting, 116-121, 164-167
Board of Boiler Rules, 143-145, 163
Board of Director's meeting minutes, 178-179
Boiler Code, *see* Standards
Boiler explosions, 48, 93
Boiler heads, *see* Heads, boiler
Boiler plate, 41, 50, 129
Boiler plate, marine, *see* Marine boiler plate
Boiler testing, 129-133
Brandywine Iron Works and Nail Factory, 38, 40, 59, 61

Bockmann, John, 7, 9, 93
Brooke, Charles, 44
Brown, John, 118
Bureau of Standards, 171
Business communication, 7, 8, 66-67, 72, 92
Byrd, William, 17

Cambria Steel Co., 142
Car record books, 82-84
Carbon copying, 111, 113, 162, 163
Carmiencke, John Hermann, 25-26
Carnegie Steel Co., 142
Carpenter, Prof. R. C., 143
Centennial celebration, Lukens, 146-148
Chandler, Alfred, 63
Chaucer, 10
Chemical analysis, 89, 110, 113, 136-138, 173
Chemical laboratory, 173
Chirographic communication, 16, 189
Christie, Joseph, 100
Civil War, 47, 76, 154
Coates, Moses, 39
Coatesville, 39, 45, 47-48
Coatesville Boiler Works, 47
Codorus, 42, 64
Colored pencil, 64, 86, 99
Commercial books, 25
Comptroller, *see* Office of the Comptroller
Consulting, 177, 180-183
Corliss Steam Engine Company, 47, 77
Corporate communication, *see* Business communication
Cornell, 142
Correspondence, 60, 66, 67, 159, 183
Columbia School of Mines, 163
Coupon, boiler plate, 171
Ctesibius, 16

Da Vinci, Leonardo, 10
Daybooks, 21, 60, 63-64
Debit and credit accounting, 64
Defective records, 86, 89-91
Defects, steel, 81, 86, 93, 113, 134, 168
Delaware River Bridge, 156
Densmore, Christopher, 64
Dew, Charles B., 20
Dictation, 112, 160
Dies, 53, 191
DiOrio, Eugene, 56
Director's meeting minutes, 178-179
Discourse communities, *see* Social discourse communities
Discursive practices, 5
Drafting linen, 119, 127
Drawing, 7, 53, 96, 97, 99, 113, 116, 119, 123-128, 150, 164
Drawing index, 119, 121-122
Dunleavy, 180

Eagleton, Terry, 10
Edgar Thompson Plant, 181
Educational texts, 34
Elgar, John, 42
Emery, Clayton, 100
Employees, *see* Mill positions
Engineering and Mining Journal, 31-32
Engraving, 25
Erskine, Robert, 19
Estes, Inc., L. V., 179, 182-183
Explosions, *see* Boiler explosions
Extensometer, 95

Fairbanks & Ewing, 94
Federal Slitting Mill, 38
Federal Trade Commission, 154
Ferguson, Eugene, 116, 123, 127
Filing systems, 104
Financial record keeping, *see* Accounting writing
Fireboxes, locomotive, 156-157
Flanging, 123, 128, 191
Forbes, E. A., 181
Ford, Bacon & Davis, 183-185
Forms, *see* Printed forms
Foucault, Michel, 45, 104, 189
Franklin, Benjamin, 10
Franklin Institute, 50, 75, 93, 95, 127, 142
Fritz, John, 146-148
Furnace journals, *see* Journals, furnace

General Electric, 129
Geological surveys, 24-25
Gibbons, Abraham, 46, 59, 64, 67
Gibbons, Martha Lukens, 46, 64
Goodfellow, Alfred, 110, 113, 118, 179, 181
Gordon, Frank H., 53, 99, 180
Graphic arts, *see* Drawing
Graphs, 136, 138, 165, 169
Greek Empire, 15
Group work, 5, 7, 8, 143, 165, 167, 183, 190

Hagley Museum and Library, 50, 54, 104
Haines, R. B., 98
Haller, Cynthia, 4
Haldeman, P. C., 110, 181
Hamilton, William H., 110, 113, 165, 179
Hargis, Charles, 10
Hasenclever, Peter, 19
Heads, boiler, 50, 66, 69, 120, 123, 192
Heine Safety Boiler Works, 143
Hermelin, Samuel Gustaf, 18
Hibernia Iron Works, 44
Holly, Alexander, 78
Howe, Henry, 163, 172
Humpton, Charles F., 100
Humpton, Howard, 51, 100
Humpton, Joseph, 51, 53, 54, 77, 99, 100, 178
Humpton, May, 51, 100
Humpton, William G., 134
Hunt, Robert W., 142
Huston, Abram Francis, 46, 50-51, 54, 77, 99, 100, 146-147, 157-158, 177, 180, 183
Huston, Charles Lukens, 46, 50-51, 76-77, 95-97, 99, 101-102, 104, 110, 111, 127, 134, 138-141, 155, 157, 158, 160-161, 168, 177, 178, 180, 183
Huston, Dr. Charles, 46, 51, 59, 67, 77, 87, 93, 95, 168
Huston, Isabella Lukens, 46
Huston, Charles Lukens III, 46
Huston, Charles Lukens Jr., 46
Hydraulic presses, 53

Illustrations, in publications, 25, 27, 28
Indicator cards, 129-131, 192
Index, drawing, 119, 121-122
Ingots, 51, 78, 108, 110, 112, 136, 192
Ingot molds, 51, 81, 110-112, 136
Inspection, 77, 86, 93, 113, 134, 135, 139, 171
Intraplant communication, 111-112
Inventories, 92
Iron Act of 1750, 38
Iron Trade Review, 31
Ivins, William, 165

Jobs, *see* Mill Positions
Jones & Laughlin, 142
Johnson Iron Works, 120
Journals, furnace, 20, 21, 59, 60, 63-64, 82
Journal of the Franklin Institute, 50, 75, 93

Kersey, Jesse, 39
Kitchell, William, 25-26
Knowledge exchange, creation and transfer, 4-5, 16, 20-21, 25, 27, 76, 122, 129, 165, 181, 189

Laboratory report, 116
Lake Erie Boilerworks, 143
Lawrence, Anna, 100
Ledgers, 21, 60, 63-64
Letter writing, *see also* Correspondence, 59, 66-67, 107, 186
Letterbooks, 2, 8, 37-38, 45-46, 60, 67, 70-72, 67, 96, 98, 101-104, 111-112
Letterpress books, *see* Letterbooks
Literature, 5-6, 10
Locomotive fireboxes, *see* Firebox, locomotive
Lukens centennial celebration, *see* Centennial celebration, Lukens
Lukens, Dr. Charles, 39-42
Lukens, Rebecca, 38, 42-46, 59, 66, 154
Lukens, Solomon, 44, 61

Machine shop cards, 116, 117
Manhole heads and saddles, 50, 149
Management consulting, *see* Consulting
Mann's Parchment Copying Paper, 70-72
Manufacturing Board meeting minutes, 179-180
Marine boiler plate, 113, 134, 154, 170-171
Martha Furnace Diary and Journal, 21
Martin, H. G., 113, 138
Martin Ford, Mrs. Anna E., 51, 100
M. I. T., 116, 143
Massachusetts State Board of Boiler Inspection, 143-144
Master Mechanics Association, 143
McKenna, Christopher, 182
Meetings, 178-180
Meeting minutes, *see* Director's, Manufacturing and Operations meeting minutes
Meinholtz, H. G., 143
Metallurgy, 93, 143, 173
Microphotographs, 136
Mill positions, 92, 98, 100, 109, 156-158
Miller, Benjamin, 178
Miller, C. F., 143
Miller, Carolyn, 9
Molds, *see* Ingot molds
Moore, Emmett, 123
Moyers, H. C., 113, 114, 115, 125-126

National Tube Co., 142
Newspapers, trade, 28
Nock, George, 21, 24

Office of the Comptroller, 182
Open hearth process, 78-81, 110, 112, 191
Open hearth heat books, 86
Operations Committee meeting minutes, 178-180
Otis Steel Co., 101, 104, 181
Overman, Frederick, 25, 28-29

Patents, 95, 97
Patterns, wooden, 118, 192
Payroll records, 92, 158
Pencoyd Iron Works, 129
Pennock, Isaac, 38-41
Penrose, Charles, 46
Persian Empire, 15-16
Philadelphia and Columbia Railroad, 44-45, 61
Philadelphia Lancaster Turnpike, 39
Philadelphia Ranger Boiler, 47
Philo of Byzantium, 15
Photography, 50, 136
Pit furnaces, soaking, 53, 81, 108-110, 191
Pittsburgh-plus pricing, 154
Plate straightening machine, 53, 77, 97
Pollard and Redgrave, 6
Prediscursive communication, 4, 19, 21, 25, 59, 72, 77, 162, 167, 189
Price card, 174-175
Principio Iron Works, 18, 21
Printed forms and reports, 112, 114, 116, 130, 132
Printing techniques, 25
Proceedings of the American Society for Testing Materials (ASTM), 138-141, 172
Product guide, 146, 149
Professional associations, 26, 27
Public Relations, 143, 146-148, 173
Publishing industry, 25
Puddle mill books, 66, 82, 85
Puddling 47, 66, 76, 77

Quakerism, 37, 66

Railroads, Coatesville, 44-48, 61, 76
Railroads, plant, 82, 113
Railway Gazette, 143
Raymond, Rossiter, 31
Record keeping systems, 75
Red pencil, *see* Colored pencil
Rejections, 113-114, 116, 141, 164, 171
Renaissance, 6, 15-16, 25
Reports, general, 163-164
Report of Plates Rejected, 116

Report of Tests of Steel, 86, 113, 114, 116, 136, 137, 173
Riley, Vince, 161
Robertson, Miss Helen, 104, 162, 178-181
Roller dishing device, *see also* Spinning machine, 123, 127, 128
Rolling mill, description, 78, 110, 192
Roman Empire, 15
Rosenblatt, Louise, 10
Russell, Frank, 100

Scheffler, Judith, 56
Schlaudecker, S. D., 182-183
Scientific discourse, 8, 9
Scientific testing, 48, 93, 95, 129, 134, 136, 167, 173
Segregation in steel, 138-141
Shorthand, 120, 122
Schulman, Joel, 10
Simmons-Boardman Publishing Company, 34
Sized linen, *see* Drafting linen
Skaggs, Julian, 56, 66, 108
Social discourse communities, 1, 3-6, 37, 66, 95-96, 104, 107, 129, 141-143, 145, 159, 171, 189
Societies, professional, *see* Professional associations
Spackman, Horace B., 53, 77, 99, 100, 158-161, 178, 179, 180
Spackman, Owen, 100
Spinning machine, 50, 123
Spotswood, Alexander, 18
Standardized forms, *see* Printed forms
Stamped templates, 96, 98
Standards, 50, 93, 108, 133, 141-145, 167, 170, 171
State reports, 24-25
Steamboat inspections, *see also* Inspection, 93, 170
Steel plate, 50, 53, 110, 138, 170
Stenographer typist, 9, 104, 112, 150, 160, 162, 183
Stratton, S. W., 171
Swedenborg, Emanuel, 18
Swedish industrial spying, 18
Straightening rolls, *see* Plate straightening machine
Suffern & Son, 181
Systems, the Magazine of Business, 173, 176, 183

Tacit knowledge, 18, 20, 31, 114, 122
Taggert, Howard, 113, 133, 141, 143, 178
Tebeaux, Elizabeth, 6, 21, 25
Tensile strength, 47, 77, 93, 129, 133, 138, 141, 170
Testing, 48, 77, 89, 93, 107, 111, 113, 127, 129-133, 162, 172, 173, 176
Testing machine, 48, 77, 93, 94, 176
Testing personnel, 172-173
Thomas, David, 142
Tonnage, in pay, 82
Tonnage records, 86, 87, 88, 179
Trade journals, 28, 143
Transactions of the American Institute of Mining Engineers, 30
Transactions of the American society of Mechanical Engineers, 116
Transcribed discussions, 31
Triadelphia Iron Works, 47,
Tuball Iron Works, 18
Typewriting, 76, 99, 101
Typist, *see* Stenographer typist

United Engineering & Foundry, 53, 107, 154, 155, 163
United States Board of Supervising Inspectors of Steam Vessels, 170
United States Department of Commerce, 172
United States Patent Office, 96, 97
United States Steel, 154

Valley Iron Works, 47
Vannen, Irwin, 100
Van Ormer, J. R., 99
Viaduct and Welded Steel Shapes, 47
Visual communication, *see* Drawing

Waste books, 21
Watertown Arsenal, 172
Webster, William R., 141
Welding Society, 181
West Point, 116
Wiley Technical Series, 34
Wilmington and Northern Railroad, 47, 76
Wolcott, Robert, 159, 180, 183
Woodhull, F. H., 113
World War I, 53, 154, 178
Workers, *see* Mill positions
Worth Brothers, 47
Woodcuts, 25, 27

Yates, JoAnne, 4, 7, 67, 72, 179, 183

In Praise

This meticulously researched and well-documented scholarly analysis of technical communication in a major American steel company is written in a language accessible to the layperson. Johnson provides a wealth of visual documentation and opens up the equally impressive data resource hidden in the levels of business and technical communication within the artifacts that contain the language of this work. In Lukens Steel she has found the perfect institution to demonstrate how much more we can learn when we include the artifacts of technical communications in our studies of industries. This book will appeal to industrial archaeologists, historic preservationists, avocationals and professionals interested in the iron industry, and readers fascinated by the uses of language. A superb and readable study!

Edward J. Lenik
Registered Professional Archaeologist

We technical writers seem to think that our field started with the software industry or, if we have very long memories, with Grace Hopper's efforts for the U.S. Navy in the 1970s. However, as Carol Johnson points out in *The Language of Work,* technical writing has a much longer and richer history than that. By the 1800s, as apprenticeship systems with oral traditions gave way to mechanized systems run by professionals, information had to be written down. However, writing books that readers could understand and navigate easily didn't come automatically. *The Language of Wor*k shows how our predecessors eventually turned logs and notes into standardized texts and industry bibles, creating many of the types of information design that we use today.

Susan Fowler
Fast Consulting
Member of the Society for Technical Communication

SELECTED TITLES FROM

Baywood's Technical Communications Series

Series Editor, Charles H. Sides

OUTSOURCING TECHNICAL COMMUNICATION
Issues, Policies and Practices
Edited by Barry L. Thatcher and Carlos Evia

MOTIVES FOR METAPHOR IN SCIENTIFIC AND TECHNICAL COMMUNICATION
Timothy D. Giles

TOGETHER WITH TECHNOLOGY
Writing Review, Enculturation, and Technological Mediation
Jason Swarts

RESOURCES IN TECHNICAL COMMUNICATION
Outcomes and Approaches
Editor: Cynthia L. Selfe

COMMUNICATIVE PRACTICES IN WORKPLACES AND THE PROFESSIONS
Cultural Perspectives on the Regulation of Discourse and Organizations
Editors: Mark Zachry and Charlotte Thralls

INTERNSHIPS
Theory and Practice
Editors: Charles H. Sides and Ann Mrvica

FREEDOM OF INFORMATION IN A POST 9-11 WORLD
Editor: Charles H. Sides

CROSS-CULTURAL COMMUNICATION
Perspectives in Theory and Practice
Thomas L. Warren

For details on these titles from Baywood's Technical Communications Series, visit http://baywood.com.

ONLINE EDUCATION
Global Questions, Local Answers
Kelli Cargile Cook and Keith Grant-Davie

— WINNER OF THE 2006 NCTE AWARD FOR BEST COLLECTION OF ESSAYS —
IN TECHNICAL OR SCIENTIFIC COMMUNICATION

POWER AND LEGITIMACY IN TECHNICAL COMMUNICATION
Volume I: The Historical and Contemporary Struggle for Professional Status
Editors: Teresa Kynell-Hunt and Gerald J. Savage

POWER AND LEGITIMACY IN TECHNICAL COMMUNICATION
Volume II: Strategies for Professional Status
Editors: Teresa Kynell-Hunt and Gerald J. Savage

TWISTED RAILS, SUNKEN SHIPS
The Rhetoric of Nineteenth Century Steamboat and Railroad Accident Investigation Reports, 1833-1879
R. John Brockmann

VISUALIZING TECHNICAL INFORMATION
A Cultural Critique
Lee E. Brasseur

EXPLODING STEAMBOATS, SENATE DEBATES AND TECHNICAL REPORTS
The Convergence of Technology, Politics and Rhetoric in the Steamboat Bill of 1838
R. John Brockmann

EXPLORING THE RHETORIC OF INTERNATIONAL PROFESSIONAL COMMUNICATION
An Agenda for Teachers and Researchers
Editors: Carl R. Lovitt and Dixie Goswami

THE EMERGENCE OF A TRADITION
Technical Writing in the English Renaissance, 1475-1640
Elizabeth Tebeaux

PUBLICATIONS MANAGEMENT
Essays for Professional Communicators
Editors: O. Jane Allen and Lynn H. Deming